Springer Tracts in Modern Physics
Volume 134

Springer-Verlag Berlin Heidelberg GmbH

Springer Tracts in Modern Physics

Volumes 118–134 are listed at the end of the book

Covering reviews with emphasis on the fields of Elementary Particle Physics, Solid-State Physics, Complex Systems, and Fundamental Astrophysics

Manuscripts for publication should be addressed to the editor mainly responsible for the field concerned:

Gerhard Höhler
Institut für Theoretische Teilchenphysik
Universität Karlsruhe
Postfach 6980
D-76128 Karlsruhe
Germany
Fax: +49 (7 21) 37 07 26
Phone: +49 (7 21) 6 08 33 75
Email: hoehler@fphvax.physik.uni-karlsruhe.de

Johann Kühn
Institut für Theoretische Teilchenphysik
Universität Karlsruhe
Postfach 6980
D-76128 Karlsruhe
Germany
Fax: +49 (7 21) 37 07 26
Phone: +49 (7 21) 6 08 33 72
Email: johann.kuehn@physik.uni-karlsruhe.de

Thomas Müller
IEKP
Fakultät für Physik
Universität Karlsruhe
Postfach 6980
D-76128 Karlsruhe
Germany
Fax:+49 (7 21) 6 07 26 21
Phone: +49 (7 21) 6 08 35 24
Email: mullerth@vxcern.cern.ch

Roberto Peccei
Department of Physics
University of California, Los Angeles
405 Hilgard Avenue
Los Angeles, California 90024-1547
USA
Fax: +1 310 825 9368
Phone: +1 310 825 1042
Email: robertop@college.ucla.edu

Frank Steiner
Abteilung für Theoretische Physik
Universität Ulm
Albert-Einstein-Allee 11
D-89069 Ulm
Germany
Fax: +49 (7 31) 5 02 29 24
Phone: +49 (7 31) 5 02 29 10
Email: steiner@physik.uni-ulm.de

Joachim Trümper
Max-Planck-Institut
für Extraterrestrische Physik
Postfach 1603
D-85740 Garching
Germany
Fax: +49 (89) 32 99 35 69
Phone: +49 (89) 32 99 35 59
Email: jtrumper@mpe-garching.mpg.de

Peter Wölfle
Institut für Theorie
der Kondensierten Materie
Universität Karlsruhe
Postfach 69 80
D-76128 Karlsruhe
Germany
Fax: +49 (7 21) 69 81 50
Phone: +49 (7 21) 6 08 35 90/33 67
Email: woelfle@tkm.physik.uni-karlsruhe.de

Joachim Wosnitza

Fermi Surfaces of Low-Dimensional Organic Metals and Superconductors

With 88 Figures

 Springer

Dr. Joachim Wosnitza

Physikalisches Institut
Universität Karlsruhe
Engesserstraße 7
D-76128 Karlsruhe

Cataloging-in-Publication Data applied for

Die Deutsche Bibliothek - CIP-Einheitsaufnahme

Wosnitza, Joachim:
Fermi surfaces of low dimensional organic metals and
superconductors / Joachim Wosnitza.

(Springer tracts in modern physics ; Vol. 134)
ISBN 978-3-662-14845-7   ISBN 978-3-540-49238-2 (eBook)
DOI 10.1007/978-3-540-49238-2

NE: GT

Physics and Astronomy Classification Scheme (PACS):
71.25.Hc, 74.70.Kn, 75.30.Fv

ISBN 978-3-662-14845-7

Typesetting: Camera-ready copy from the author using a Springer T$_E$X macro-package
Cover design: Springer-Verlag, Design & Production
SPIN: 10511697   56/3144-5 4 3 2 1 0 - Printed on acid-free paper

Preface

The activities in the synthesis and investigation of organic metals were originally initiated three decades ago by the proposal of polymeric superconductors with extraordinary high transition temperatures. The extensive research finally succeeded in the discovery of highly conducting organic metals that showed superconductivity. However, this superconductivity was at rather low temperatures and in the beginning occurred only under high pressure. Although the mechanism responsible for the superconductivity in the organic compounds known to date has nothing to do with the originally proposed excitonic interaction, the physical properties of these low-dimensional materials show a fascinating variety of unusual and exciting effects.

The metallic and superconducting states of these organic compounds are highly sensitive to external parameters such as the pressure, magnetic field, and cooling rate. By proper treatment the metallic state may be suppressed *and charge* or spin density wave states, antiferromagnetism or weak ferromagnetism, can easily be induced or destroyed. Sometimes even a coexistence of these contradictory states appears to be possible.

A key for a better understanding of organic metals is the knowledge of the electronic structure, i. e., the Fermi surface. The most direct and, therefore, most important tools for the experimental determination of the Fermi surface are the de Haas–van Alphen (dHvA) and Shubnikov–de Haas (SdH) effects. These methods, which were already fully developed in the late 1950s, have undergone a great revival in recent years, stimulated in particular through the success in resolving the Fermi-surface topology in organic superconductors.

In this review an introduction to the structural, electronic, and superconducting properties and a comprehensive overview of the present-day knowledge of the Fermi surfaces of organic low-dimensional metals will be given. Thereby, attention is focused mainly on the quasi two-dimensional charge transfer salts, whereas the quasi one-dimensional organic metals are reviewed in a somewhat shorter fashion. The recently discovered three-dimensional fullerenes based on C_{60} are excluded altogether in this overview, especially because to date only very limited experimental data concerning the Fermi surfaces are available.

The material is presented in five chapters. After the mainly historical background given in the introduction in Chap. 1 an outline of the basic structural, electronic, and superconducting properties is presented in Chap. 2. Chapter 3 gives a brief introduction to the theoretical understanding of magnetic quantum oscillations and angular dependent magnetoresistance oscillations. The experimental results are discussed in detail in Chap. 4. Finally, Chap. 5 gives some concluding remarks and an outlook of future developments.

New results and new materials are appearing constantly in this highly topical field of the "fermiology of organic superconductors and metals". Therefore, this review can give only a picture of the subject as it currently is. In the forthcoming years many of the present-day questions may be solved, whereas certainly other new aspects of solid-state properties might occur.

Finally it is a pleasure to thank the following people who have contributed to this work in different ways: first of all, Prof. H. von Löhneysen, whose continuous interest, encouragement and discussions motivated the work considerably; Prof. G. Crabtree, Drs. U. Welp, and W. Kwok for initiating the interest in the experimental technique of dHvA measurements, for their sharing of experimental knowledge, and for many fruitful discussions; Prof. J. Williams, Drs. H. Wang, D. Carlson, and U. Geiser for sharing their crystals and their knowledge about organic metals; Prof. D. Schweitzer for his spontaneous and continuous support with high-quality crystals and for many discussions; Drs. N. D. Kushch, H. Müller, and J. Schlueter for the supply of crystals; Drs. W. Biberacher, J. S. Brooks, M. Dressel, T. Ishiguro, M. Kartsovnik, M. Kund, M. Lang, V. N. Laukhin, T. Sasaki, and Y. Sushko for many useful discussions; N. Herrmann, D. Beckmann, S. Wanka, Dr. G. Goll, and Dr. X. Liu for their large contributions to many of the experimental results presented here; Profs. E. Dormann and P. Wölfle for valuable suggestions improving the final state of this review; finally the many colleagues who allowed the publication of their results and supplied information prior to publication.

Karlsruhe, January 1996   *Joachim Wosnitza*

Contents

1. Introduction

Usually organic materials are regarded as being prototypes of electrical insulators. In everyday life a vast set of different carbon-based substances, such as fibers, coatings, synthetic rubbers, plastic bags and so on, surround us. However, some of the organic compounds showed at least at room temperature metallic conductivity, i. e., decreasing resistivity with decreasing temperature. These "synthetic metals", however, usually have conductivities orders of magnitude less than Cu. On the other hand, the technology in the synthesis of organic materials is well developed and there is a unique possibility in tailoring solid state compounds with the desired mechanical, electrical and even magnetic properties. However, it appears that the interactions governing these properties are very subtle and to date we are still far from achieving the "molecular engineering" of substances with desired electronic and magnetic properties.

The first organic compounds which showed electronic conductivity at room temperature were semiconductors [1]. Much effort was spent on the synthesis of materials with higher conductivity. An important step in the development of organic metals was the discovery of TTF–TCNQ [2] consisting of the organic electron acceptor tetracyano-p-quinodimethane (TCNQ) and the electron donor tetrathiafulvalene (TTF). This material exhibits a metallic conductivity, σ, along the stacking direction with $\sigma \approx 10^4 \, \mathrm{S \, cm^{-1}}$ near $60 \, \mathrm{K}$ (Cu has $\sim 10^6 \, \mathrm{S \, cm^{-1}}$ at room temperature). Below this temperature, however, a Peierls metal–insulator transition [3] occurs and the conduction electrons no longer contribute to the conductivity. As will be discussed in more detail in Sect. 2.2.3, Fröhlich has pointed out that a one-dimensional (1D) metal in principle is unstable against a lattice distortion and a concomitant gap opening at the Fermi level [4]. Indeed, in many quasi-1D organic conductors the ground state is characterized by collective ordering phenomena like charge density wave (CDW) or spin density wave (SDW) states [5]. Numerous investigations deal with organic systems showing these Peierls transitions in order to get a better understanding of the nature of this low-temperature ground state. Apart from the 1D materials with TTF other charge transfer salts based on so-called arenes, such as pyrene, perylene, or fluoranthene [6], are well-studied Peierls systems.

The 1D and possibly even 2D metals are also especially interesting, namely the theoretically predicted occurrence of a Luttinger liquid [7]. For example, the experimental search for the expected behavior of a Luttinger liquid peaked after the finding of quasi 1D TTF–TCNQ. Recently, photoemission data of a 1D and some 2D organic metals revealed one of the features of a Luttinger liquid, namely the missing of a sharp Fermi edge [8, 9, 10]. Up to now, however, it is not clear whether the low-dimensional organic conductors can be described within the Luttinger picture.

Since the beginning of the investigations of organic metals the search for superconductivity was one of the main goals. This was especially stimulated by a hypothesis made by Little [11] that in organic polymers with highly polarizable side chains the pairing of electrons to Cooper pairs should be highly favorable. This, he argued, could result in extremely high superconducting transition temperatures because the region of positive charge generating the attractive interaction is caused by the displacement of an electron with much smaller mass than that of the moving ions constituting a lattice vibration in inorganic superconductors. However, as should be stated here clearly, in spite of the large efforts spent so far no superconductivity based on the exitonic mechanism proposed by Little has yet been found.

The first organic superconductor discovered in 1979, 15 years after Little's stimulating idea, was $(TMTSF)_2PF_6$ [12], where TMTSF stands for tetramethyltetraselenafulvalene. At ambient pressure, however, this material shows a metal–insulator transition to a low temperature state which has been recognized as a SDW phase. Only with the application of the considerably large pressure of $\sim 12\,kbar$ can this phase be suppressed and superconductivity with a transition temperature of $T_c \approx 0.9\,K$ occur. Shortly afterwards a series of isostructural superconductors obtained simply by replacing the charge-compensating anion PF_6 by AsF_6, SbF_6, TaF_6, ReO_4, FSO_3, and ClO_4 were discovered [13]. Of all these materials, however, only $(TMTSF)_2ClO_4$ is a superconductor ($T_c \approx 1.4\,K$) at ambient pressure [14].

Apart from these so-called Bechgaard salts based on TMTSF, a large variety of organic superconductors with different building blocks were found within the next few years. Figure 1.1 shows the principal molecules of more than 60 different organic superconductors known to date [13]. The largest number of superconductors is based on the molecule BEDT–TTF (bisethylenedithio-tetrathiafulvalene or ET for short). The first superconducting material based on ET was $(ET)_2(ReO_4)_2$ [15]. This compound also needs pressure (4.5 kbar) to suppress an insulating state and allow superconductivity at $T_c \approx 2\,K$. The organic charge transfer salt with the highest T_c to date (besides the fullerenes) is also based on the ET molecule. κ-$(ET)_2Cu[N(CN)_2]Cl$ has a very complicated low-temperature phase diagram but shows superconductivity at $T_c \approx 12.8\,K$ with only moderate pressure of 0.3 kbar [16]. So far at least 30 different ET-based superconductors are known, many of them at ambient pressure with transition temperatures up to ~ 11.6 K.

In contrast to the Bechgaard salts which are quasi-1D, the ET materials are characterized by their two-dimensional (2D) electronic structure. In both salts the electronic bands are formed by the overlapping molecular π orbitals. Along certain directions the distances between either Se–Se or S–S is less than the van der Waals radii of 3.96 Å or 3.6 Å, respectively. Within the ET compounds many different ways of stacking of the ET molecules is possible. These polymorphic phases are denoted by α, β, κ, θ, etc. and will be discussed in Sect. 2.3.1.

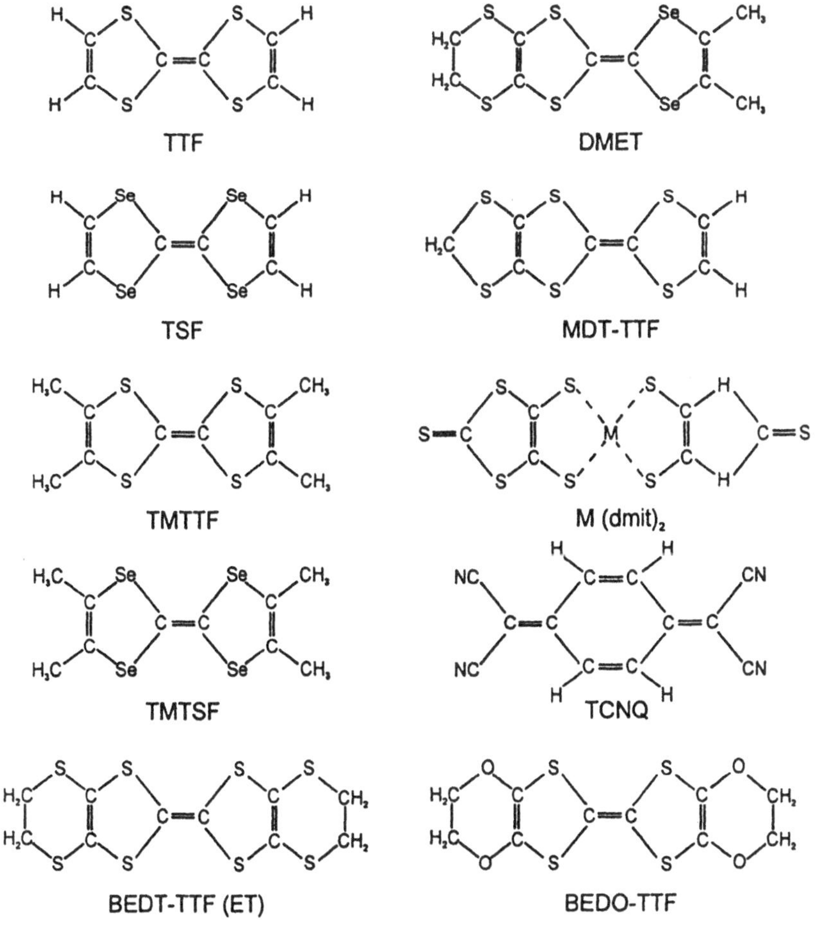

Fig. 1.1. Molecular structures of some building blocks in organic conductors and superconductors

Other organic superconductors are based on DMET (dimethyl-ethylene-dithio-diselenedithiafulvalene), an asymmetric molecule hybridized from ET and TMTSF [17]. According to its origin the physical properties are somewhere between one and two dimensional. Also based on asymmetric donors are the ambient-pressure superconductors $(MDT-TTF)_2AuI_2$ with $T_c \approx 3.5\,K$ [18] and $(DMET-TSeF)_2X$ with $X = AuI_2$ and I_3 (T_c below $1\,K$), where MDT–TTF stands for methylenedithio-TTF and DMET–TSeF for dimethyl-ethylenedithio-tetraselenafulvalene [19].

According to the BCS theory T_c is to a first approximation proportional to the Debye temperature $\Theta_D \propto \mu^{-1/2}$, where μ might be related to the reduced molecular weight. Therefore, a main goal was the reduction of the molecular weight of the organic constituents. Indeed, two organic superconductors based on the molecule BEDO–TTF (bisethylenedioxy-TTF), where the outer S atoms in ET are substituted by O, have been synthesized. Contrary to expectations, however, the transition temperatures are relatively low around $1\,K$ [20, 21].

Furthermore, superconductivity has been found in some salts based on the acceptor molecule $M(dmit)_2$, where $M =$ Ni, Pd, or Pt and dmit is 4,5-dimercapto-1,3-dithiole-2-thione ($= C_3S_5$) [22, 23].

The properties of organic metals both in the superconducting and the metallic states are quite unusual. The low-temperature state is highly sensitive to external parameters like pressure, field or cooling rate. This manifests itself in very rich phase diagrams, where superconductivity, CDW, SDW and even ferromagnetism exist next to each other. Many of the organic superconductors show an extraordinarily large change of T_c with pressure. Some of the synthetic metals have ground states under ambient conditions which can be easily modified under moderate pressure. This results, for example, in two largely different superconducting transition temperatures in β-$(ET)_2I_3$ or under certain conditions in reentrant superconductivity in κ-$(ET)_2Cu[N(CN)_2]Cl$.

To date the nature of the superconducting state in organic metals is controversial and either BCS or unconventional behavior has been suggested. Experimental data are not yet conclusive and are interpreted in both ways. The missing of the Hebel-Slichter peak at T_c and the observation of a peak at lower temperatures in some NMR experiments were interpreted as a sign of unusual behavior. This and the existence of antiferromagnetic ordering and superconductivity close to each other in TMTSF and also in some ET salts are sometimes taken to suggest an electron pairing mechanism due to spin fluctuations [24, 25]. On the other hand, tunneling, magnetization, and specific-heat data are consistent with the usual electron–phonon coupling with a tendency, however, towards strong coupling. Since the low-dimensional organic superconductors behave in many aspects similar to the layered cuprate high-T_c superconductors the same mechanism principle might be responsible for the occurrence of superconductivity. In the latter materials some experimental evidence points to a non-Fermi liquid-like behavior with, however, a well-defined Fermi surface. To account for this apparent contradiction, therefore, the concept of spin-charge separation resulting in a Luttinger liquid has been proposed [26]. However, the actual relevance of the Luttinger-liquid picture to 2D organic superconductors is unclear and remains controversial. Many other mechanisms for superconductivity in organic metals, such as electron-molecular vibration interaction, so-called "g-ology" [27], excitonic models, and so on, have been discussed [25]. An excellent introduction to these different theories and to the physics of organic superconductors is given by Ishiguro and Yamaji [28].

One important point understanding metallic and superconducting behavior is that of the Fermi surface (FS). The key tools for mapping out the FS topology are measurements of the de Haas–van Alphen (dHvA) and Shubnikov–de Haas (SdH) effect. Of course, these magnetic quantum oscillations are only observable if a closed extremal orbit of the charge carriers in k space exists. Therefore, in 1D material with only open FS sheets the observation of dHvA or SdH oscillations is not possible. However, in the quasi 1D organic metals $(TMTSF)_2X$, where X stands for the charge-compensating anion, other unexpected effects in high magnetic fields of different origin have been observed. These were, for example, the field-induced SDW state, giant angular dependent magnetoresistance oscillations (AMRO), and the so-called rapid oscillations. These findings have stimulated enormous efforts both experimentally and theoretically, where new models have been developed clarifying some of the observed peculiarities. To date, however, many open questions still exist, challenging both experimental and theoretical studies to clarify the remaining problems. In Sects. 2.2 and 4.1 some of these basic features of the 1D organic metals will be reviewed.

The 2D organic conductors, on the other hand, are ideally suited for the observation of both dHvA and SdH effects. With the increasing quality of the samples first SdH oscillations in κ-$(ET)_2Cu(NCS)_2$ [29] and β-$(ET)_2IBr_2$ [30] were discovered in 1988. In the following years magnetic quantum oscillations in many other organic metals were found [31]. These investigations made a considerable contribution to understanding the electronic properties of organic charge transfer salts. For many compounds it was possible to map out the exact topology of the FS which allowed a direct comparison with the predictions of band-structure calculations. Very often a remarkable agreement between experiment and the calculated FS was found. In these calculations the FS was obtained by the tight-binding Hückel method under rather crude approximations.

All the ET compounds are characterized by their nearly perfect 2D band structure resulting in an almost cylindrical form of the FS. For some materials it was possible to determine quantitatively the transfer integral between the ET layers, i. e., the dispersion of the energy-momentum relation perpendicular to the highly conducting planes, the so-called warping. In some cases an extremely 2D band structure was found which resulted in very unusual features such as giant amplitudes and a strong harmonic content of the magnetic quantum oscillations. Other ET salts with additional 1D open FS topologies have been found to be unstable against a SDW transition. This leads to peculiar, unexpected and still not quite understood behavior in SdH and dHvA measurements.

Both 1D and 2D organic metals are, of course, three-dimensional crystals with a certain degree of three-dimensionality also in the electronic system. This results in a corrugated form of the FS which gives rise to a distinct kind of AMRO. Therefore, measurements of the angular dependence of the resistance in magnetic fields has become a new and powerful tool to determine principal features of low-dimensional band structures. Section 3.3 will give a brief introduction into the present understanding of the AMRO in quasi 1D and 2D metals.

In recent years successful measurements of magnetic quantum oscillations have been employed to uncover the nature of the low-temperature states in organic metals. SdH measurements under pressure show the changes of the band-structure due to the increasing dimensionality. The observation of dHvA oscillations in the superconducting state was possible so far in one ET salt [32] and gave more insight into the kind of scattering mechanism of the quasiparticles in the Shubnikov phase.

From the temperature dependence of the dHvA (or SdH) oscillations it is possible to extract the effective cyclotron mass (see Sect. 3.1). Comparisons of the mass obtained by these measurements with values from band-structure calculations, cyclotron resonance and specific-heat measurements are sometimes inconsistent. Whether strong electron–electron or electron–phonon interactions play the dominant role for this discrepancy is still under considerable debate and further studies have to deal with this question. Chapter 4 will review the present-day knowledge of the highly active field of "the fermiology of organic superconductors".

2. Some Principal Properties

2.1 Synthesis

Several review articles and a comprehensive book already exist which describe the detailed synthesis of the miscellaneous organic donor molecules and the experimental procedures to obtain organic conducting salts [33]. Therefore, only a few introductory remarks concerning the principal crystal growth technique will be given here.

The usual method of producing high-quality single crystals is a process called "electrocrystallization", during which the organic electron donor molecules are oxidized electrochemically. In the same process crystals are built with the charge-balancing anions. Typically, an H-configuration cell with an ultrafine porosity glass frit and two platinum electrodes is used for the crystal growing. The donor molecules, e. g., TMTSF or ET with an appropriate solvent, are put inside the anode compartment. A supporting electrolyte, e. g., NBu_4X, is added both to the anode and cathode compartment, where Bu is n-butyl and X is the desired monovalent anion such as ClO_4^- or $Cu(NCS)_2^-$. Electrocrystallization is a slow process which takes typically from one week up to a few months. The current density is kept at the lowest possible level ($\sim 1\,\mu A/cm^2$) to initiate crystal growth. After seed crystals appear the current density is reduced further to avoid the generation of divalent donor cations and to obtain large single crystals. Sometimes after harvesting the first batch of crystals a second and third growth process in used solution results in higher quality crystals. Usually, the crystals obtained are black. The quasi 1D materials have a needle-like form with the long axis of a few millimeter length being the highly conducting one. The 2D salts have plate-like shapes of a few millimeter side length and smaller thickness perpendicular to the highly conducting planes of a few tenths of a millimeter.

To obtain the many different phases of the metals based on ET donor molecules, the exact current density, the right solvent, and the correctly prepared starting material are essential. The ET-I_3 system, for example, has more than a dozen different phases with different stoichiometry [34]. For the usual charge transfer with the formula $(ET)_2I_3$ six phases are known, some of which can be converted from the crystals already grown by a special heat and pressure treatment.

Considerable effort is currently being spent on the improvement of crystal qualities and the evaluation of specific growth conditions to obtain the reliable crystallographic phases required. Consequently, the number of new organic metals is growing steadily. Often for many of these salts only the structural parameters and a few physical properties are investigated. Superconducting species, however, are attracting greater attention and are studied more carefully.

2.2 Quasi One-Dimensional Systems

This section gives a brief description of the crystal structure, the electronic properties and the superconductivity of the so-called Bechgaard salts with the general formula $(TMTSF)_2X$, where X represents a monovalent anion [35]. $(TMTSF)_2PF_6$ was the first organic compound which showed superconductivity [12]. This discovery has boosted an enormous amount of activity, resulting in a whole family of superconducting compounds all based on TMTSF. The isomorphous $(TMTTF)_2X$ salts, on the other hand, do not show superconductivity.[1] The building block tetramethyltetrathiafulvalene differs from TMTSF only in the substitution of S atoms instead of the four Se atoms (Fig. 1.1). The sulfur-based charge transfer salts, however, show a shallow minimum in the resistivity at $\sim 200\,K$ and are semiconducting at lower temperatures [37]. Besides their intensively investigated and sometimes unusual superconducting properties the 1D organic materials served as model systems for the study of spin-Peierls and SDW transitions. Since in the (P, T) phase diagram these transitions occur in the vicinity of superconductivity it is speculated that magnetic ordering is related to the mechanism for superconductivity.

2.2.1 Crystal Structure of $(TMTSF)_2X$

The structure of tetramethyltetraselenafulvalene $[=(CH_3)_4C_6Se_4]$ is shown in Fig. 1.1. TMTSF is a planar brick-like molecule. In the charge transfer salts $(TMTSF)_2X$ these bricks are stacked in columns with a slight tendency towards dimerization. As an example, two views of the crystal structure of $(TMTSF)_2PF_6$ are shown in Fig. 2.1 [38]. The columns are formed along the a axis which is also the axis with the highest electrical conductivity. The anions, here PF_6^-, are located between these columns. The crystal symmetry is triclinic with space group $P\bar{1}$. The lattice parameters at room temperature for the cell containing one chemical formula unit (Z=1) are $a = 7.297\,\text{Å}$, $b = 7.711\,\text{Å}$, $c = 13.522\,\text{Å}$, $\alpha = 83.39°$, $\beta = 86.27°$, and $\gamma = 71.01°$.

---

[1] Recently some indications for superconductivity in $(TMTTF)_2Br$ have been found at pressures of 26 kbar [36]. See Sect. 2.2.3.

Two TMTSF molecules transfer one electron to PF_6^-. This monovalent electron acceptor can be replaced by AsF_6^-, SbF_6^-, and TaF_6^-, all with octahedral symmetry and superconducting transition temperatures in the compound of $\sim 1\,K$ under applied pressures of $\sim 10\,kbar$. The anions with tetra-

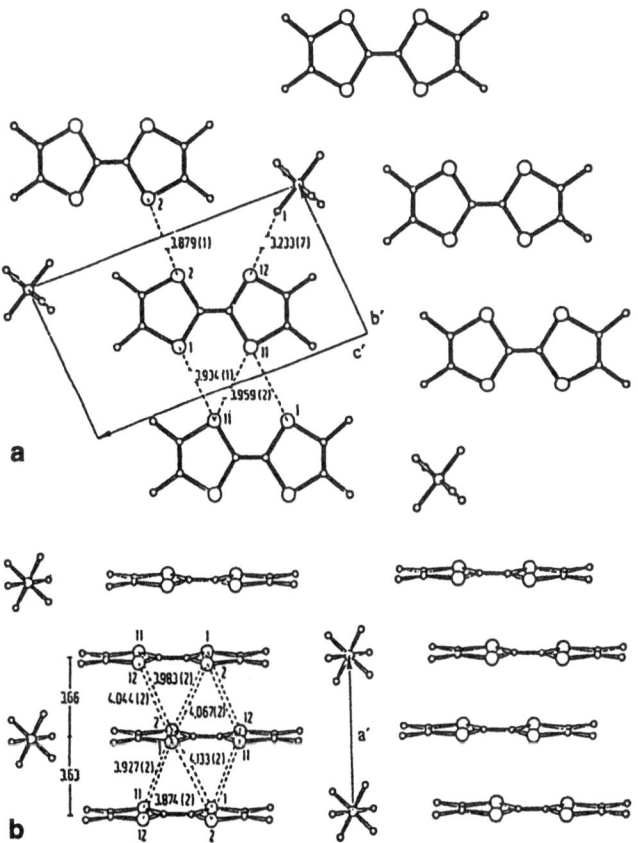

Fig. 2.1. (a) Crystal structure of $(TMTSF)_2PF_6$ viewed along the a direction and (b) $10°$ tilted side view of the TMTSF stacks. a', b', and c' are the projections of a, b, and c. The distances are in Å. From [38]

hedral symmetry, ClO_4^- and ReO_4^-, keep the crystal structure isomorphous. $(TMTSF)_2ClO_4$ is the only quasi-1D metal which becomes superconducting under ambient pressure [14]. The acceptor X can further be replaced by a large variety of other anions some containing even magnetic constituents such as $FeCl_4^-$ [39]. However, the only other superconducting Bechgaard salt found to date is $(TMTSF)_2FSO_3$ with the highest T_c of this family at $\sim 3\,K$ [40]. Salts containing the electron donor TMTTF form isomorphous structures but become insulating at lower temperatures. The compounds $(TMTTF)_2X$ are, therefore, well suited for comparative studies between isostructural superconductors and insulators. Section 2.2.3 discusses the reasons which are believed to be responsible for the electronic differences observed.

2.2.2 Electronic Structure

In Fig. 2.1 the most important intermolecular Se–Se distances are given [38]. As can be seen from this figure, the spacings of the Se atoms are around 3.9 Å both along the stacking direction and in the perpendicular plane. This value is approximately the sum of the van der Waals radii of the Se atoms, 3.96 Å. Although the distances within and perpendicular to the stacks are almost the same in $(TMTSF)_2X$ the overlap is strongest within the columns. This is due to the fact that between two neighboring planar TMTSF molecules only two Se atoms are in side-by-side contact. In the stacking direction, on the other hand, four Se orbitals are overlapping. In addition, the overlap is made by π-electron orbitals which extend in the stacking direction. This kind of overlap enables large electron transfers along the stacks, less coupling along the b direction, and the weakest transfer integrals along the c direction where the anions and the terminal methyl groups act as a barrier.

The calculation of the exact band structure from first principles, however, is rather complex and requires considerable simplifications. The usual and very successful method to calculate the band structure of organic charge transfer salts is a tight-binding method, called extended Hückel approximation. In this approximation, one starts from the molecular orbitals (MO) which are approximated by linear combinations of the constituent atomic orbitals. Each MO can be occupied by two electrons with antiparallel spins. These valence electrons are assumed to be spread over the whole molecule. Usually, only the highest occupied molecular orbital (HOMO) and the lowest unoccupied molecular orbital (LUMO) are relevant and are, therefore, considered in most band-structure calculations [41].

The principal properties of the low-dimensional organic metals, however, can be sketched already by the simple free-electron approximation, although here the delocalization of the π electrons is overestimated. Without any interaction in one dimension the electron energy levels are just given by $\epsilon(k) = (\hbar^2/2m)k_a^2$, where m is the free-electron mass and k_a is the electron wave vector in the direction of highest conductivity. This dispersion relation in the repeated zone scheme is shown by the dashed lines in Fig. 2.2a. Inclusion of the correction of the free-electron parabola due to Bragg reflection at $\pm\pi/a$ gives the dispersion depicted as the solid lines in Fig. 2.2a. The available electrons have to be filled into the possible energy levels. Since in $(TMTSF)_2X$ each TMTSF has transferred half an electron to the anion X the highest band is filled to three-quarters. This results in a Fermi energy ϵ_F within the HOMO and the material should behave metallicly. Indeed, the $(TMTSF)_2X$ salts are metallic at room temperature. The overlap of the electronic wave functions in the other directions is very weak and, therefore, the energy $\epsilon(\boldsymbol{k})$ is nearly independent of k_b and k_c. Figure 2.2b shows schematically the resultant 3D FS which consists of two almost parallel sheets perpendicular to k_a. Without any overlap along the b and c direction the sheets would be exactly parallel as shown by the dashed lines.

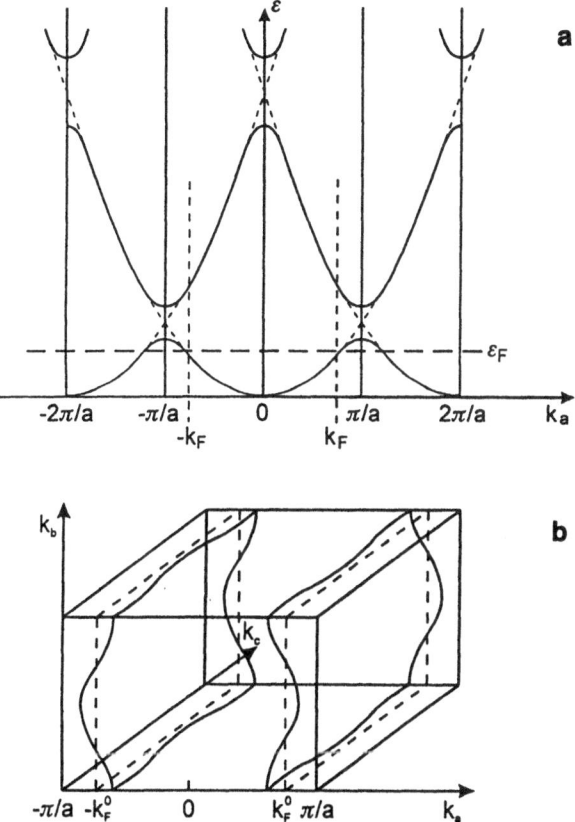

Fig. 2.2. (a) Schematic illustration of the free-electron-like 1D dispersion relation. ϵ_F is the Fermi energy, k_F is the Fermi wave vector, and a is the lattice constant. (b) 3D schematic view of the resulting Fermi surface. Dashed line without interstack overlaps. Solid lines with small transfer integrals

In the tight-binding band approximation the analogous result for the dispersion relation can be written as

$$\epsilon(\boldsymbol{k}) = 2t_a \cos(\boldsymbol{a}_m \boldsymbol{k}) + 2t_b \cos(\boldsymbol{b}_m \boldsymbol{k}) + 2t_c \cos(\boldsymbol{c}_m \boldsymbol{k}), \qquad (2.1)$$

where $\boldsymbol{a}_m$, $\boldsymbol{b}_m$, and $\boldsymbol{c}_m$ are intermolecular distances in the crystal lattice directions a, b, and c, $\boldsymbol{k}$ is the electron wave vector, and t_i is the electron transfer energy along the i direction. For $(\mathrm{TMTSF})_2 X$ both $\boldsymbol{b}_m$ and $\boldsymbol{c}_m$ correspond to the lattice constants b and c, while $\boldsymbol{a}_m = a/2$ due to the dimerization of the TMTSF molecules (see Fig. 2.1). From plasma frequency measurements

[42] the transfer energies are estimated to $t_a \approx 0.28\,\text{eV}$ and $t_b \approx 0.022\,\text{eV}$. The transfer along the c direction is of the order $1\,\text{meV}$.

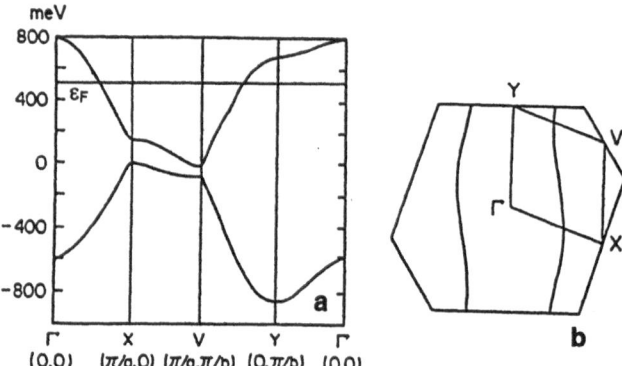

Fig. 2.3. (a) Calculated band structure and (b) Fermi surface of $(\text{TMTSF})_2\text{AsF}_6$ representative for all $(\text{TMTSF})_2X$. From [43]

Numerical tight-binding band-structure calculations result in the approximative dispersion relation which is valid in the neighborhood of the FS [43]

$$\epsilon(\boldsymbol{k}) \simeq 2[t_\text{I}\cos(\boldsymbol{k}\boldsymbol{b}) \pm t_\text{S}\cos(\tfrac{1}{2}\boldsymbol{k}\boldsymbol{a})], \qquad (2.2)$$

where t_S and t_I are averaged transfer energies within the stacks and approximated interstack interactions in b direction, respectively. Because of the dimerization two bands ($\pm$) occur in the calculation. For $(\text{TMTSF})_2\text{PF}_6$ the calculation results in $t_\text{S} = 0.38\,\text{eV}$ and $t_\text{I} = 0.024\,\text{eV}$. Therefore, in a first approximation the electronic band structure can be regarded as quasi 1D. The remaining dispersion along b, however, becomes important for charge and spin density wave phase transitions. Figure 2.3 shows the calculated band structure and the 2D FS of the first Brillouin zone of $(\text{TMTSF})_2\text{AsF}_6$. For different anions the dispersion relation and the FS are almost indistinguishable [43]. Note the similarity of the dispersion relation along $\varGamma X$ with the text-book free-electron picture of Fig. 2.2.

2.2.3 Ground-State Instability

Although the electronic system in the Bechgaard salt is quasi 1D the crystal lattice of course is three dimensional. Due to the strong anisotropy, however, the lattice has vibrational modes of very different energies. In addition, because of the weak coupling the organic crystals are not very rigid. This results, for example, in the peculiar so-called mechanical kink effect. By carefully pushing a point on the side of a thin needle-like crystal of $(\text{TMTSF})_2\text{ClO}_4$ a pair of kinks along the long axis of the crystal can be produced [44]. These kinks can be removed and even moved along the needle axis by appropriate local pressure treatment.

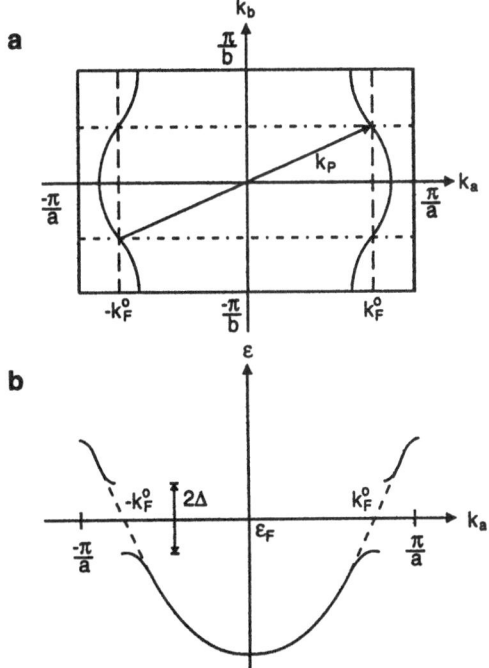

Fig. 2.4. (a) Schematic representation of a quasi 1D Fermi surface with nesting vector $k_P = (2k_F, \pi/b, 0)$. The dash-dotted line is the resulting new Brillouin zone. (b) Opening of the Peierls gap 2Δ at $\pm k_F$ in the dispersion relation

The strong mutual interaction between the electronic system and the lattice of the Bechgaard salts is often seen in a combined lattice and electronic phase transition, the so-called Peierls instability. Metals with a pronounced 1D character are unstable against a perturbation with wave number $2k_F$, where k_F is the Fermi wave vector. Figure 2.4a shows schematically the 2D projection of a quasi 1D metallic FS with a cos-like modulation along k_b (see also Fig. 2.2b). As can be seen in Fig. 2.4a, the left sheet of the FS can be superposed on the right sheet of the FS by the translational vector k_P. In other words it is said that the FS is nested by the vector k_P. If now the lattice (or more generally an external field) modifies its structure with the nesting or Peierls wave vector, k_P, a new Brillouin zone is generated (dashed lines in Fig. 2.4a) with k_P being exactly a reciprocal lattice vector. At this new zone boundary the energy of the electron levels will be reduced due to Bragg reflection, as shown schematically in Fig. 2.4b. For the case of complete FS nesting the whole FS disappears in the ground state. If some parts of the FS have different curvatures, i.e., higher dimensionality, the nesting will be incomplete and parts of the FS will remain, eventually even as new closed parts.

Depending on the relative energies for the lattice distortion and the electron-energy redistribution the Peierls transition will occur. The critical temperature, T_P, can be calculated and is given by

$$k_B T_P = 1.13 \, \epsilon_B \exp\left(-\frac{A_P}{\lambda^2}\right), \qquad (2.3)$$

where k_B is the Boltzmann constant, ϵ_B specifies the energy region where the electron distribution is perturbed ($\gg k_B T$), λ is the electron–phonon coupling constant, and A_P is a material constant depending on the Fermi velocity, lattice constants and normal-mode frequencies. Below T_P, therefore, together with a lattice change an energy gap of 2Δ at ϵ_F will occur and the quasi 1D metal will become an insulator. Concomitant with the lattice modulation an electron density modulation occurs. As can be seen from the tangential form of the electron dispersion at the new Brillouin zone boundary, the electronic density of states is drastically enhanced at the wave vector $2k_F$. This transition, therefore, is often also called charge density wave (CDW) transition [45].

As first pointed out by Overhauser [46], the perturbing potential for the opening of a gap at the FS might also come from a spin redistribution. This so-called spin density wave (SDW) orders the spins of the itinerant electrons antiferromagnetically with the same nesting vector k_P as for the CDW. Below the corresponding transition temperature, T_M, the metal becomes insulating and an analogous gap at the FS is formed [47].

The occurrence of the energy gap 2Δ below the Peierls transition temperature allows in principle the collective motion of the electrons under the influence of an applied electric field. This holds as long as the energy $\hbar k_F v_{Fr}$ of the moving electrons is less than Δ, where v_{Fr} is the velocity of the collectively moving electrons. However, this so-called Fröhlich mode [4] is very sensitive to lattice imperfections because it is a true 1D movement.

For the case where the bandwidth or the warping, i.e., the transfer integral (t_b, respectively t_I in (2.2)), is small the Coulomb repulsion between the electrons becomes important. A limited screening of the electron charge in a narrow band due to restricted electron movement can lead to a localized electron lattice, a so-called Wigner crystal. This, in fact, has been observed in the strongly 1D material TTF–TCNQ where in addition to the $2k_F$ Peierls lattice distortion a $4k_F$ modulation was found [48, 49, 50]. The estimated value for the on-site Coulomb repulsion U in TTF–TCNQ is $U/4t_b \simeq 0.9$ extracted from the frequency dependence of the NMR relaxation time [51] and the susceptibility above the Peierls transition [52].

Many of the Bechgaard salts show at ambient pressure a metal–insulator transition, T_{MI}, around a few tens of K as can be seen from the resistivity behavior shown for several $(TMTSF)_2X$ salts in Fig. 2.5 [35]. This transition is

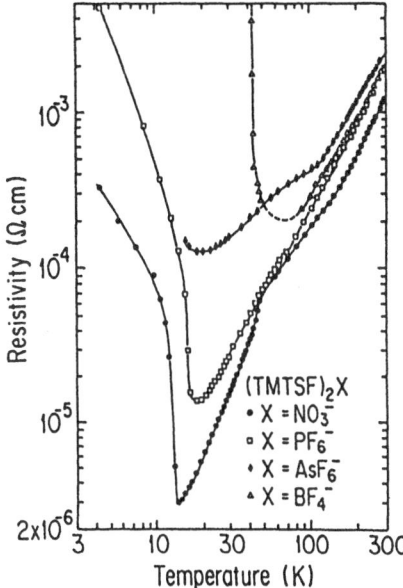

Fig. 2.5. Temperature dependence of the dc resistivity for various $(TMTSF)_2X$ salts in double-logarithmic scale. From [35]

neither accompanied by a lattice distortion nor by an isotropic decrease of the magnetic susceptibility. The latter effect should occur for a Peierls transition with a CDW, since then the carrier density should decrease exponentially with the opening of the gap Δ.

Measurements of the magnetic susceptibility, χ, below T_{MI} show a strongly anisotropic behavior. This can be seen in Fig. 2.6 where as an example χ of $(TMTSF)_2AsF_6$ is plotted [53, 54, 55]. For B parallel to a and c^* ($=$ the direction perpendicular to both a and b) χ behaves similarly with a small anomaly around T_{MI} and a slight increase below the transition. For B parallel b' ($=$ perpendicular to a and c^*) χ vanishes exponentially below T_{MI} (see inset of Fig. 2.6). This behavior is characteristic of an antiferromagnetically ordered state where the spin orientation is alternating along the b' direction. This kind of spin ordering is confirmed for $(TMTSF)_2AsF_6$ and also for $(TMTSF)_2PF_6$ both by ESR and NMR experiments [53, 56, 57, 58, 59]. The metal–insulator transition in $(TMTSF)_2X$ is, therefore, believed to be a SDW ordering. With ^{1}H NMR experiments it was even possible to estimate the SDW nesting vector $Q = (Q_a, Q_b, Q_c)$, with $Q_a = 2k_F$, $Q_b \simeq 0.2b^*$, and $Q_c \simeq 0$, and the amplitude $\sigma \approx 0.08\mu_B$ of the SDW modulation, where μ_B is the Bohr magneton [60, 61]. The value of σ less than one means that the moments are not well localized but retain still an itinerant character. The vector Q is in good agreement with the nesting vector of the calculated band structure (see Fig. 2.3).

The large variety of ground states in quasi-1D organic conductors not only depends on the chemical composition of the organic molecule and the

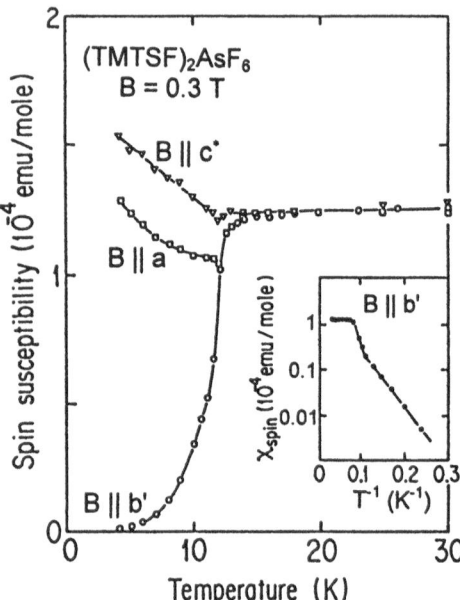

Fig. 2.6. Temperature dependence of the static spin susceptibility for (TMTSF)$_2$AsF$_6$. The inset shows the exponential decrease of the susceptibility in the b' direction below T_{MI}. From [55]

inorganic anion but can also be changed by the application of pressure. Figure 2.7 shows a generic phase diagram for TM$_2$X salts, where TM stands for TMTTF or TMTSF [62]. At the left side of the diagram (point a) salts like (TMTTF)$_2$PF$_6$ and (TMTTF)$_2$AsF$_6$ are located. At ambient pressure these salts show a maximum conductivity around 230 K. Below this temperature a Mott–Hubbard state with Coulomb correlations as in TTF–TCNQ and a Wigner-like localization are thought to be the reason for the resistivity increase [63] (not shown here). At lower temperatures (≈ 15 K) the systems show a spin-Peierls (SP) transition, where the antiferromagnetically correlated spins condense into a singlet state with a combined lattice distortion. This was elucidated by X-ray diffraction [64], magnetization [65], and ESR measurements [66]. At pressures above ~ 13 kbar a SDW emerges as the new ground state [67]. Since hydrostatic pressure increases the electronic overlap, i. e., the transfer integrals, this demonstrates that the SDW state is more stable than the SP state with increasing dimensionality. The salt (TMTTF)$_2$Br is already in the SDW ground state at ambient pressure and low temperatures (point b in the phase diagram) [68]. In principle, under high enough pressure it can be expected that the SDW is suppressed and superconductivity might appear. Indeed, under a pressure of approximately 25 kbar early experiments showed a sharp resistance decrease near 3.5 K in some samples [69]. In a recent experiment, however, at 24 kbar a sudden increase of the resistivity at 5 K due to the SDW transition was found. At 26 kbar, finally, a magnetic-field dependent resistance drop at $T_c \approx 0.8\,\mathrm{K}$ could be observed [36]. Although the resistance is not completely zero and the transition is rather broad the independence of the $R(T)$ behavior on the measuring current and the increase of R in magnetic fields is interpreted as a confirmation of bulk superconductivity.

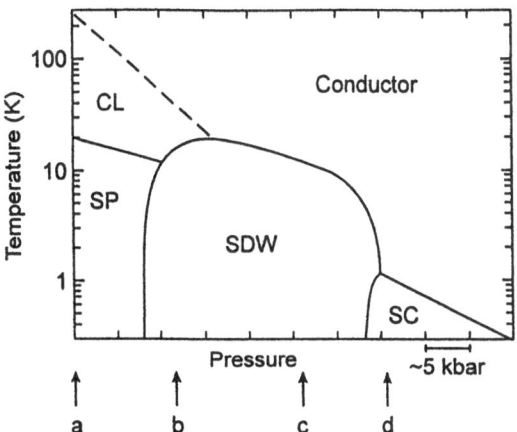

Fig. 2.7. Generalized phase diagram for the TM_2X salts. The different states mean charge localized (CL), spin-Peierls (SP), spin density wave (SDW), and superconducting (SC). The arrows indicate the ambient pressure location of some compounds described in the text. From [62]

In the TMTSF salts which are believed to be of higher dimensionality than the TMTTF analogues the SDW can be suppressed under applied pressure thereby stabilizing the superconducting phase. $(TMTSF)_2PF_6$, for example, becomes superconducting at a pressure of $\sim 12\,\mathrm{kbar}$ [12]. All $(TMTSF)_2X$ compounds with anions of octahedral symmetry show similar phase diagrams with somewhat differing critical pressure for superconductivity (point c in the phase diagram). $(TMTSF)_2ClO_4$ is the only member of this family which has "high enough dimensionality" to show superconductivity under ambient pressure (point d) [14].

If, however, the latter compound is cooled rapidly ($\geq 50\,\mathrm{K/min}$) a metal–insulator transition around $6.5\,\mathrm{K}$ occurs and no superconductivity is visible. The anion ClO_4 has tetrahedral symmetry and, therefore, two orientations with respect to the surrounding molecules are possible. If the crystals are cooled slowly around $\sim 24\,\mathrm{K}$ the anions order and form a superlattice. This was verified by the observation of an X-ray diffraction peak at $Q = (0, 1/2, 0)$ [70]. Rapid cooling, however, prevents this ordering and the anions are frozen in random directions.

Other TMTSF salts with non-centrosymmetric anions also show this anion ordering process at varying temperatures [71]. However, these compounds do not become superconducting at low temperatures and ambient pressure.

Under applied pressure a glassy behavior of the anion order is sometimes observed and often superconductivity is stabilized.

Another fascinating phenomenon in the quasi 1D materials was found soon after the discovery of superconductivity under the influence of applied magnetic field, B. In the initial experiment [72] the magnetoresistance of $(TMTSF)_2PF_6$ in the metallic state for pressures larger than approximately 6.5 kbar showed structures proportional to $1/B$ when the field was applied parallel to c, the direction of lowest conductivity. A periodicity in $1/B$ is reminiscent of the Shubnikov–de Haas (SdH) effect (see Sect. 3.2) and was first, indeed, interpreted in this way. To observe SdH oscillations, however, the existence of a closed electron orbit on the FS is a necessary condition.[2] In the TMTSF salts both from band-structure calculations and other experimental facts the occurrence of a closed FS is hard to envisage. Because of the low frequency of the periodic signal it was suggested that due to an imperfect nesting of the FS in a density-wave state small parts of the FS remain. Some other facts, however, excluded this proposal definitely. First, the oscillations only appear above a certain threshold field and, second, the frequency of the oscillation is temperature dependent. To enhance the mystery of the low-temperature high-field state of the Bechgaard salts further, a little later steps in the Hall resistivity were discovered in $(TMTSF)_2ClO_4$ [73]. This was the first observation of the quantum Hall effect (QHE) in a bulk crystal.[3]

Initiated by these unexpected but highly stimulating results a theory now commonly known as the "standard model" for field induced spin density waves (FISDW) was developed which explains the main features of the observed peculiarities. After the initial work of Gor'kov and Lebed' [75] many other groups have worked out the concept sometimes with refined models [76] which are excellently described in more detail in Ref. [28].

The principal effect of the applied magnetic field is an effective reduction of the dimensionality of the electronic system. For most Bechgaard salts the dimensionality has to be increased by the appropriate pressure at the outset. If, however, the compound is brought into the metallic state the electronic system is highly sensitive to other external parameters. A magnetic field causes the electrons to move in trajectories along the open orbits with an oscillation in the perpendicular direction of a finite width which becomes increasingly more 1D as the field is increased. As discussed above, a 1D system is inherently unstable against a DW distortion. Therefore, the magnetic field may induce a transition into a SDW state, the so-called FISDW.

In a simple model the quasi 1D dispersion in the metallic state can be written as

---

[2] Under certain conditions in high enough fields magnetic-breakdown orbits (see Sect. 4.1) might be possible without a closed FS. However, in the Bechgaard salts the band structure has not the appropiate topology for such kind of orbits.

[3] In contrast to the original QHE [74], however, only kinks in the resistance, ρ_{xx}, but no points where $\rho_{xx} = 0$ were observed.

$$\epsilon(\boldsymbol{k}) = \hbar v_{\mathrm{F}}(|k_x| - k_{\mathrm{F}}) + 2\,t_b \cos(k_y b) + 2\,t'_b \cos(2k_y b), \qquad (2.4)$$

where the x dispersion close to ϵ_{F} has been linearized with the Fermi velocity $v_{\mathrm{F}} = 2at_a \sin(ak_{\mathrm{F}})/\hbar$ and t'_b describes higher harmonics originating from transverse transfer interactions of the form t_b^2/t_a. The latter term describes the deviation from perfect nesting and has to be large enough to prevent the SDW transition at zero field. The transfer integrals t_c and t'_c in the k_z direction (see (2.1)) are omitted in (2.4) for simplicity. However, in the standard model a finite value of t'_c is responsible for the mentioned threshold field. t'_b and t'_c are of the order of $1\,\mathrm{meV}$ and $0.01\,\mathrm{meV}$, respectively. Within mean-field approximation Gor'kov and Lebed' calculated the spin susceptibility $\chi(\boldsymbol{Q})$ in an applied magnetic field B for a SDW wave vector $\boldsymbol{Q_0} = (2k_{\mathrm{F}}, \pi/b, \pi/c)$. From the divergence of $\chi(\boldsymbol{Q_0})$, which is equivalent to the appearance of a SDW state, they obtained for $|t'_c| \ll |t'_b|$ a condition for the field-induced SDW transition temperature, T_{SDW}:

$$A \simeq J_0^2(2t'_b/\hbar v_{\mathrm{F}} \kappa_{\mathrm{M}}) \ln\left(\frac{T_0}{T_{\mathrm{SDW}}}\right), \qquad (2.5)$$

where A and T_0 are constants, J_0 is the Bessel function of zeroth order, and $\kappa_{\mathrm{M}} = beB/\hbar$ is the magnetic length or the period of the electron oscillation. Later, this approach was extended for more general SDW wave vectors [76]. From (2.5) one obtains finite transition temperatures T_{SDW} except for the values of B where J_0 is zero. Since for argument $z \geq 1$ the Bessel function $J_0(z)$ can be approximated by $J_0(z) = (2/\pi z)^{1/2} \cos(z - \pi/4)$, the zero points are given by the condition $z = \pi(n + 3/4)$ with integers n. Therefore, different SDW states are passed periodically in $1/B$ giving rise to the observed structures [72]. This model can also explain the QHE in $(\mathrm{TMTSF})_2\mathrm{PF}_6$ [77] because in SDW states no extended states at ϵ_{F} are present [78].

Figure 2.8 shows the principal phase diagram predicted by the standard model together with a recent experimental phase diagram of $(\mathrm{TMTSF})_2\mathrm{PF}_6$ [79]. The very good confirmation of the theory is obvious. There seems to be, however, substantial disagreement between the standard model and experimental results for $(\mathrm{TMTSF})_2\mathrm{ClO}_4$. At high fields ($\mu_0 H > 26\,\mathrm{T}$) a reentrant behavior, i. e., the recovery of the metallic, respectively semiconducting, state was discovered as shown by the dashed line in Fig. 2.8a [80, 81] (see also Fig. 2.9b). This would be a serious discrepancy between theory and experiment since the $n = 0$ state should be the most stable state in an applied magnetic field. Theoretically, several explanations for the reentrance line have been proposed which suggest that band-structure details, the anion ordering, or a pseudogap due to fluctuations in high magnetic fields might be responsible for the observations [82, 83, 84, 85]. However, in a recent experiment the phase diagram of $(\mathrm{TMTSF})_2\mathrm{ClO}_4$ has been revised substantially [90]. A distinctive influence of the anion ordering on the DW transition has been suggested as being responsible for the extracted phase diagram (see below).

In a completely different model it is proposed that the superconducting pairing interaction is the essential ingredient to observe the FISDW. In this theory the reentrance to a final metallic state at sufficiently high fields is predicted [86]. Indeed, some additional experimental evidence in agreement with these predictions was found in (TMTSF)$_2$NO$_3$, a system neither showing superconductivity nor FISDW [87]. Contrary to the theoretical prediction, however, recent magnetoresistance data were interpreted as a possible indication for a FISDW phase also in this non-superconductor [88]. In addition, the model of [86] has some severe numerical discrepancies and is also not capable of explaining all the experimental facts.

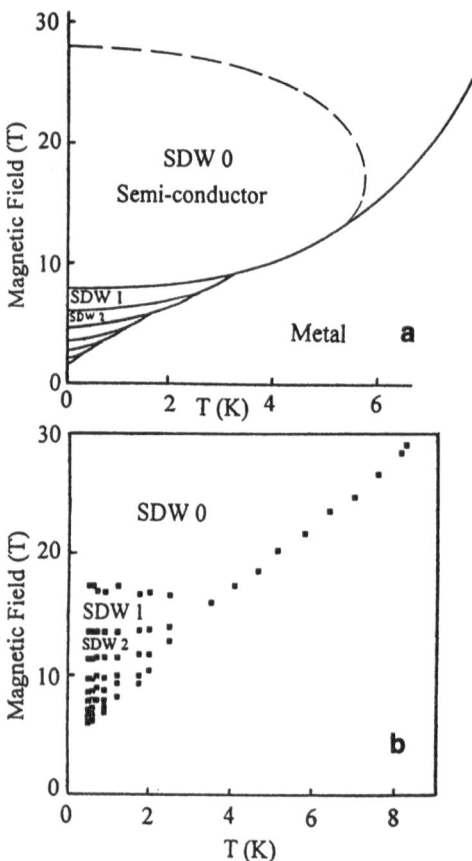

Fig. 2.8. (a) Phase diagram of the field-induced spin density wave (FISDW) states predicted by the standard model. The dashed line indicates the experimentally supposed reentrance in (TMTSF)$_2$ClO$_4$. (b) Phase diagram of (TMTSF)$_2$PF$_6$ obtained from Hall effect and magnetoresistance measurements at approximately 6 kbar. From [79]

In a recent experiment the low-temperature phase diagrams of the Bechgaard salts $(TMTSF)_2X$ with $X = PF_6$ and ClO_4 for fields up to 30 T and pressures up to 16 kbar have been mapped out systematically [89]. Figure 2.9a shows in a 3D diagram once again the satisfactory confirmation of the standard model for $X = PF_6$ up to the highest field. At higher pressures the higher field necessary to induce the SDW states is in agreement with the increased pressure-induced dimensionality. For $X = ClO_4$ (Fig. 2.9b) at ambient pressure reentrant behavior into a so-called very high field insulating

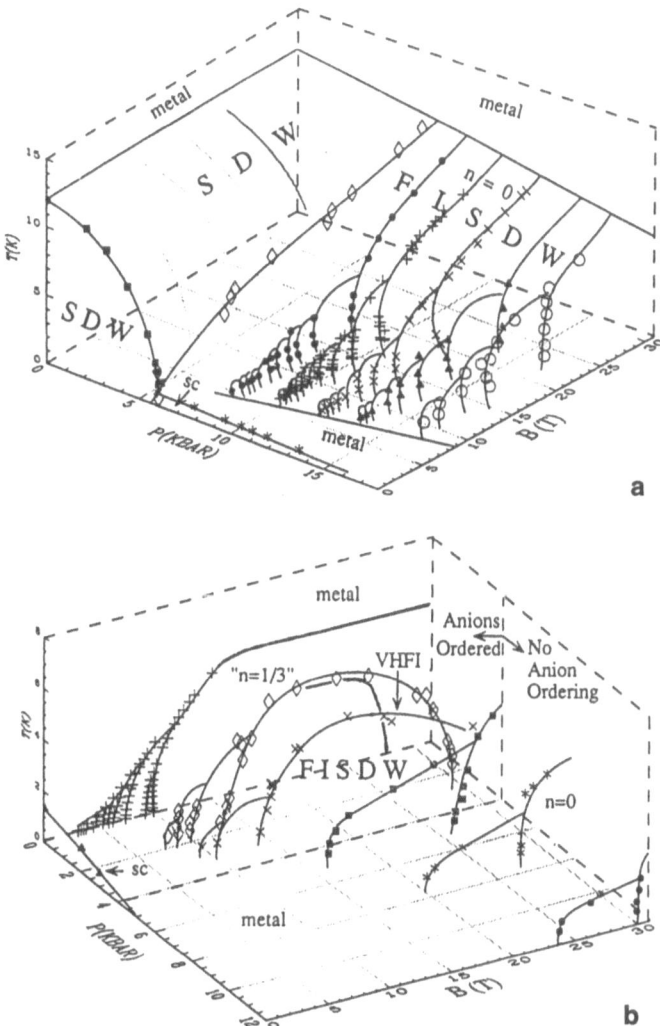

Fig. 2.9. Experimentally determined pressure-field-temperature phase diagrams for **(a)** $(TMTSF)_2PF_6$ and **(b)** $(TMTSF)_2ClO_4$. From [89]

(VHFI) state can be seen. For pressures above $\sim 4\,$kbar the anion ordering is suppressed and the standard model seems to work also for $X = ClO_4$. These results are in qualitative agreement with a model proposed by Osada et al. [84] that takes into account the anion ordering. However, for this improved standard model some points are different from experimental facts. In the experiment already mentioned [90] the phase diagram of $(TMTSF)_2ClO_4$ at zero pressure was mapped out with great care. It was found that the previously supposed reentrant line actually is a second phase line which terminates at a critical point inside the FISDW state (schematically sketched in Fig. 2.9b). The origin for these two independent phase transitions is believed to lie in the anion ordering transition at $24\,$K which dimerizes the system along the b direction. This leads to two pairs of open Fermi surface sheets. The main idea is that these two Fermi surfaces have distinct DW transitions. Whether the high-field state is the $n = 0$ FISDW state still remains to be clarified.

Another mystery unsolved to date are the so-called "rapid oscillations" seen in many Bechgaard salts [87, 91, 92, 93, 94]. These oscillations are seen in the magnetoresistance, the magnetization [93], and the sound velocity [95] and are periodic in $1/B$ with a frequency of a few $100\,$T. They are independent of temperature and largely independent of FISDW transitions. A possible explanation would be dHvA or SdH-like oscillations due to a small closed FS of approximately 3% of the first Brillouin zone. This, however, is highly unlikely since the temperature and field dependence is not like that for ordinary SdH oscillations (see Sect. 3.1) [94], they are almost not affected by the FISDW, and neither Hall effect measurements nor band-structure calculations give hints for a closed piece of FS in the ab plane. Some attempts were made to explain the rapid oscillations (here "rapid" means compared to the structures coming from the FISDW transitions) by the anion ordering [83, 84]. This, however, only explains the rapid oscillations in $(TMTSF)_2ClO_4$ in the FISDW state. In the metallic state an explanation based on the probability of electron–electron scattering resulting in an oscillating renormalized anion gap has been proposed [96]. Weak oscillations in $(TMTSF)_2PF_6$ where no anion ordering occurs may be a consequence of commensurability effects between the SDW and the crystalline lattice [97]. Finally, however, a conclusive theory for the occurrence of rapid oscillations in the Bechgaard salts is still missing. In Sect. 4.1 further details of the rapid oscillations will be discussed.

2.2.4 Superconductivity

To date, seven superconductors of the $(TMTSF)_2X$ family are known. Except for $(TMTSF)_2ClO_4$, however, all these salts need pressure above at least 5 kbar to become superconducting. Mostly the transition temperatures are around $1\,$K and T_c rapidly decreases with increasing pressure. The reason why only the $(TMTSF)_2X$ salts reveal superconductivity in contrast to

TTF–TCNQ and $(TMTTF)_2X$ compounds is attributed to the higher dimensionality in the former. Indeed, the lattice parameter in the b direction is the smallest for $(TMTSF)_2ClO_4$ resulting in the strongest transfer interaction perpendicular to the highly conducting a direction. An almost linear relation between b and the critical pressure which prevents the SDW and induces the superconducting state was found [98]. A certain degree of dimensionality, therefore, seems to be necessary to suppress the insulating ground states described in the previous section.

In $(TMTSF)_2ClO_4$ the briefly mentioned anion ordering is essential to obtain the superconducting ground state. As was first seen by NMR experiments [99], rapid cooling of the order $50\,K/min$ across the anion order temperature $(T_{AO} \approx 24\,K)$ prevents this ordering and instead of superconductivity a competing insulating SDW state is formed.

Although for many Bechgaard salts the application of pressure, P, is necessary to observe superconductivity a further increase of the pressure results in a rather rapid decrease of T_c of, e.g., $dT_c/dP \approx -8.7 \times 10^{-2}\,K/kbar$ in $(TMTSF)_2PF_6$ [100, 101]. In principle, this effect is nothing unusual (except for the magnitude) and very common in many "conventional" superconductors [102]. The effect can be explained with the BCS relation in its simplest form

$$T_c = 1.13\,\Theta_D \exp\left(-\frac{1}{\lambda}\right), \tag{2.6}$$

where Θ_D is the Debye temperature and λ is the electron–phonon coupling constant. Both Θ_D and λ depend on the phonon frequencies. The applied pressure stiffens the lattice remarkably, especially in soft organic crystals with rather weak intermolecular coupling. Therefore, Θ_D is expected to increase; λ, on the other hand, which is inversely proportional to some average phonon frequency, decreases. Altogether, due to the strong exponential λ dependence T_c is reduced.

To prove that the superconducting state really is a volume effect one of the most reliable methods are measurements of the Meissner effect. This was first done for $(TMTSF)_2PF_6$ under applied pressure [100]. After cooling the sample in the zero field and then applying a magnetic field of $0.0115\,mT$ perpendicular to the highly conducting axis a diamagnetic shielding signal of $\sim 100\%$ at $0.2\,K$ was measured. Cooling the sample in the same field gave a Meissner effect of $\sim 50\%$. This definitely verifies the superconductivity as being a true thermodynamic volume state. An increase of the field strength rapidly reduces the diamagnetic signal. This is typical for a strongly type-II superconductor with pinning. All the organic superconductors are characterized by a large Ginzburg–Landau (GL) parameter $\kappa = \lambda_L/\xi_{GL}$, where λ_L is the London penetration depth and ξ_{GL} is the GL coherence length. This parameter determines, for example, the ratio of the lower critical field, B_{c1}, where magnetic flux starts to penetrate into the superconductor, to the upper critical field, B_{c2}, where the superconductivity is destroyed totally. From the magnetization measurements of $(TMTSF)_2PF_6$ values of $B_{c1} \approx 0.5\,mT$ and B_{c2} between 50 and 100 mT are estimated.

For the highly anisotropic organic superconductors, of course, the critical fields depend also on the field direction. Most experimental data exist for $(TMTSF)_2ClO_4$ since for this material measurements at ambient pressure are possible. Figure 2.10 shows the temperature dependence of B_{c2} obtained from resistivity measurements [103]. The anisotropy is clearly seen. The extrapolated critical fields at zero temperature are $B_{c2}^a \approx 2.8\,T$, $B_{c2}^b \approx 2\,T$, and $B_{c2}^c \approx 0.16\,T$. These values are rather crude approximations with large error bars because of the extrapolations from $T/T_c = 0.4$ and the uncertainty in the experimental determination of B_{c2} from the midpoint of the resistive transition curve which may differ significantly from the true thermodynamical value (see also Sect. 2.3.3).

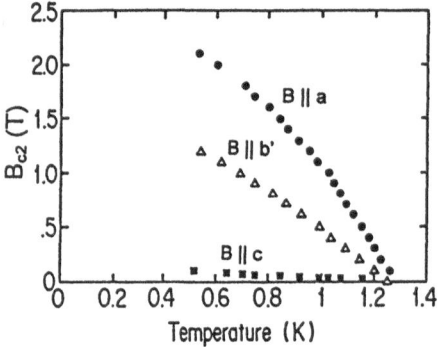

Fig. 2.10. Upper critical field, B_{c2}, vs temperature for the three principal crystal directions of $(TMTSF)_2ClO_4$. From [103]

However, for the field applied parallel to the a direction the nearly linear behavior expected for "normal" 3D superconductors is not found but, instead, a concave shape of the B_{c2} curve. This might be due to the low dimensionality of the salts and a corresponding small coherence length in a direction perpendicular to the field. With the anisotropic GL theory [104, 105, 106, 107] the critical field B_{c2} in the i direction can be written as

$$B_{c2}^i = \frac{\phi_0}{2\pi\xi_j(T) \cdot \xi_k(T)},\tag{2.7}$$

where $\phi_0 = h/2e = 2.07 \times 10^{-15}\,Tm^2$ is the flux quantum. The GL coherence length close to T_c can be approximated by $\xi_i(T) = \xi_i(0) \cdot \sqrt{T_c/(T_c - T)}$. For 3D superconductors this leads to the linear temperature dependence of B_{c2}. If, however, ξ is constant in one direction where, for instance, $\xi(0)$ is less than the lattice spacing or larger than the sample thickness, B_{c2} is proportional to $(T_c - T)^{1/2}$. This would give a behavior similar to the one observed in

Fig. 2.10 for the a direction. The calculation of $\xi_i(0)$ with (2.7) results in $\xi_a \approx 54\,\text{nm}$, $\xi_b \approx 38\,\text{nm}$, and $\xi_c \approx 3\,\text{nm}$ compared to the lattice constant $c = 1.35\,\text{nm}$. Therefore, the superconductivity in $(TMTSF)_2ClO_4$ is close to 2D but the concave shape of B_{c2}^a is not quite explicable. In addition, for the B_{c2}^b curve the same behavior should be visible.

Another explanation for the deviation of the critical field curve from linearity might be the Pauli (or Clogston) limit [108]. This is the maximum field value for the breaking of singlet Cooper pairs when the Zeeman energy exceeds the condensation energy. The Pauli limit in the absence of spin–orbit coupling[4] is given by

$$B_p = \Delta_0 \frac{1}{\sqrt{2}\mu_B},\tag{2.8}$$

where Δ_0 is the gap energy at $T = 0$. With the BCS relation $\Delta_0 = 1.76 k_B T_c$ this results in $B_p = (1.85\,\text{T/K}) \cdot T_c$. For the critical temperature of $(TMTSF)_2ClO_4$ ($T_c = 1.25\,\text{K}$) the Pauli limit is $B_p \approx 2.3\,\text{T}$. This is approximately the highest critical field observed in the experiment shown in Fig. 2.10 suggesting that the Cooper pairs in the Bechgaard salts are in the singlet state. This would be contrary to earlier assumptions of a triplet-type superconductivity [109] based on experiments which showed extremely large B_{c2} values [110]. Triplet superconductivity was believed to be caused by the close neighborhood of the antiferromagnetic to the superconducting ground state in the 1D organic metals.

The anisotropy of superconducting properties has further been seen in Meissner and shielding measurements from which B_{c1} was determined [111]. The values at $50\,\text{mK}$ for $(TMTSF)_2ClO_4$ are $B_{c1}^a \approx 0.02\,\text{mT}$, $B_{c1}^b \approx 0.1\,\text{mT}$, and $B_{c1}^c \approx 1\,\text{mT}$. Together with the values for the upper critical fields the anisotropic GL parameter κ_i can be calculated by

$$\frac{\ln \kappa_i}{2\kappa_i^2} = \frac{B_{c1}^i}{B_{c2}^i}.\tag{2.9}$$

For the above case this results in the corresponding highly anisotropic values $\kappa_a = 675$, $\kappa_b = 234$, and $\kappa_c = 15$ reflecting the very short mean free paths in the b and c directions. According to the weak coupling or short coherence length in the c direction the critical current, $j_c = 0.1\,\text{A/cm}^2$, is approximately one order of magnitude smaller than in the a direction [112].

The question whether or not the superconducting state in the Bechgaard salts is of the ordinary BCS type is still under considerable debate. A large variety of different experimental techniques have been employed to elucidate this question. One example was the measurement of the specific heat, C. The result for $(TMTSF)_2ClO_4$ is shown in Fig. 2.11 [113]. At $T_c \approx 1.22\,\text{K}$ a clear

---

[4] Spin–orbit coupling is expected to be small in most organic superconductors since the conduction bands originate from the organic molecules consisting of relatively light atoms.

anomaly in C occurs, revealing further evidence for the volume superconductivity. The apparent rounding of the jump at T_c was attributed to 1D
superconducting fluctuations but may be also due to slightly different T_c's of
the eight simultaneously measured pieces. Above T_c the specific heat follows a
fairly straight line in C/T vs T^2 as is expected for metals at low temperatures
with a linear electronic, γT, and a phononic Debye, βT^3, behavior. The values obtained in [113] are $\gamma = 10.5\,\mathrm{mJmol^{-1}K^{-2}}$ and $\beta = 11.4\,\mathrm{mJmol^{-1}K^{-4}}$
corresponding to a Debye temperature of $\Theta_D = (\frac{12}{5}\pi^4 Rn/\beta)^{1/3} = 213\,\mathrm{K}$,
where R is the gas constant and n is the number of atoms per formula unit.

The specific-heat jump at T_c is $\Delta C = 21.4\,\mathrm{mJmol^{-1}K^{-1}}$ approximated
by the equal entropy method. The BCS theory predicts for a weak-coupling
superconductor $\Delta C/\gamma T_c = 1.43$. For $(\mathrm{TMTSF})_2\mathrm{ClO}_4$ a slightly higher value
of 1.67 is found which might be an indication for strong coupling in this
material. From the exponential decrease of C below T_c an energy gap of
$2\Delta_0/k_B \approx 4\,\mathrm{K}$ has been estimated. This is in reasonable agreement with the
theoretical prediction of $2\Delta_0 = 3.5\,k_B T_c$.

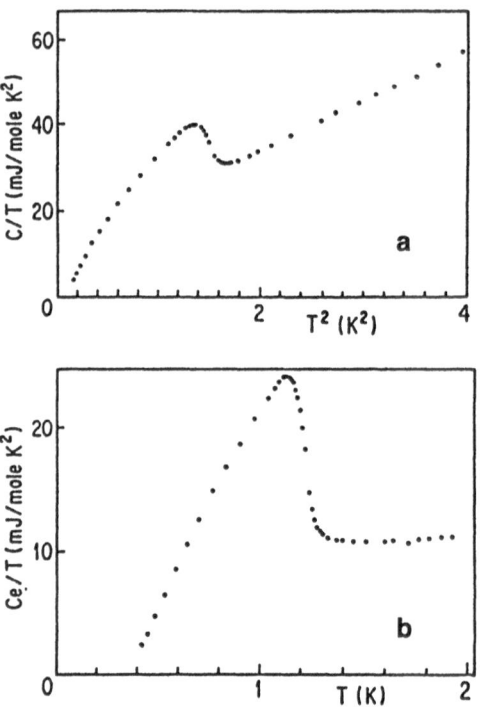

Fig. 2.11. (a) Specific heat, C, of $(\mathrm{TMTSF})_2\mathrm{ClO}_4$ plotted as C/T vs T^2.
(b) Electronic part of C after subtraction of the phononic T^3 part. From [113]

From tunneling measurements some controversial results for the energy
gap were obtained. First experiments reported an extremely large gap of

$2\Delta_0 = 3.6\,\mathrm{meV}$ for $(\mathrm{TMTSF})_2\mathrm{PF}_6$ [114] and a similar value for $(\mathrm{TMTSF})_2$-ClO_4 [115]. Later refined experiments, however, stated for $(\mathrm{TMTSF})_2\mathrm{ClO}_4$ at $0.5\,\mathrm{K}$ a much lower value of $2\Delta_0 = 0.4\,\mathrm{meV}$ for the gap in excellent agreement with the BCS relation and consistent with the above stated C value [116].

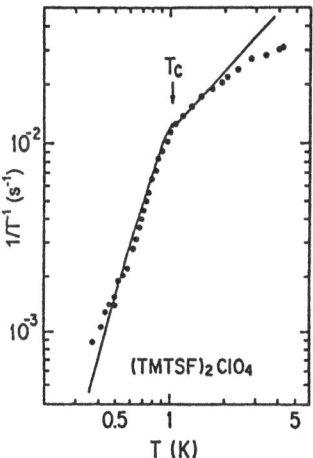

Fig. 2.12. Proton nuclear relaxation rate $1/T_1$ vs temperature of $(\mathrm{TMTSF})_2\mathrm{ClO}_4$. The solid line is a calculation with an anisotropic order parameter. From [117]

Results in contradiction with the commonly expected features of conventional superconductors were observed in proton NMR experiments. Figure 2.12 shows the temperature dependence of the reciprocal spin-lattice relaxation time, T_1, at zero field for $(\mathrm{TMTSF})_2\mathrm{ClO}_4$ [117]. The important result is that $1/T_1$ decreases roughly like T^3 just below T_c ($= 1.06\,\mathrm{K}$) without showing the typical BCS-like peak in $1/T_1$ at $T \approx 0.85\,T_c$. This so-called Hebel–Slichter peak is caused by the divergence of the density of states just at the edges of the energy gap and by the characteristic coherence factor [118]. The existence of the peak is considered as one of the crucial tests for BCS superconductivity. A possible explanation for this discrepancy might be an anisotropic order parameter which vanishes along lines on the FS. With the reasonable assumption that antiferromagnetic fluctuations mediate the attractive interaction between electrons the superconducting order parameter could have lines of zero amplitude in k space. This kind of scenario has been found, for instance, in the heavy-Fermion compound UPt_3 [119]. For the simple band structure in $(\mathrm{TMTSF})_2\mathrm{ClO}_4$ it has been shown that the order parameter can have only four different, two singlet and two triplet, symmetries. The calculated relaxation rate with the assumption $t_b \ll t_a$ for either one singlet or triplet state [120] is shown as the solid line in Fig. 2.12. The theory fits the data fairly well. Another possibility for a reduced Hebel–Slichter peak is strong scattering of the quasi-particles. This, however, seems to be highly unlikely in organic superconductors which are usually characterized by small scattering rates as is evidenced by the observed quantum effects at low temperatures.

Other, even more unconventional interpretations for the superconducting state are still the subject of controversial discussion [47]. The 1D electronic system has been claimed to be of great importance for superconductivity. The tunneling experiments already mentioned [114], infrared absorption spectra [121] revealing a magnetic field dependent feature around 30 cm^{-1}, and a decreasing thermal conductivity in the corresponding temperature range [122] were attributed to incipient superconductivity. This was interpreted as evidence for a large pseudo-gap. Therefore, it was believed that the Bechgaard salts have a high mean-field T_c which is suppressed by strong 1D fluctuations of the order parameter. However, contrary thermal conductivity data similar to the tunneling experiment also exist. In the earlier measurements [122] a decrease of the thermal conductivity below 50 K with a recovery in magnetic field was found. In contrast, later experiments showed that the thermal conductivity is steadily increasing with decreasing temperature [123]. Therefore, these results are in accordance with conventional behavior contrary to the more exotic proposals of a large pseudo-gap.

Another unconventional effect observed in the Bechgaard salts was the sensitivity of the superconducting state to nonmagnetic disorder. As mentioned above, the weak random potential due to the disordered anions in $(TMTSF)_2ClO_4$ is sufficient to destroy superconductivity. In addition, alloying experiments have revealed the strong influence of disorder on T_c [124]. Possible explanations might be triplet pairing of the superconducting carriers and also electron localization effects in these low-dimensional systems.

In conclusion, many superconducting properties of the 1D organic metals are consistent with conventional BCS behavior. However, some experiments revealed highly unusual effects which can be interpreted in controversial ways. Even more than 15 years after the discovery of the first organic superconductor these quasi-1D materials remain mysterious in many aspects. Further careful and systematic studies are necessary to resolve the open questions.

2.3 Quasi Two-Dimensional Systems

Shortly after the discovery of the superconductivity in the quasi-1D organic materials another large family showing this phenomenon was found. The organic donor molecule, where to date the most superconducting materials and the organic salt with the highest T_c (except the fullerenes based on C_{60}) were found, is bis-ethylenedithia-tetrathiafulvalene or ET (alternatively abbreviated by BEDT–TTF). Figure 2.13 shows the structure of this molecule. The stoichiometry of the ET salts can have a wide variety denoted by $(ET)_m X_n$. The valence of ET can be changed rather easily. However, most of the superconducting compounds are found in the 2:1 composition $(ET)_2 X$, where

X represents a monovalent anion. Unlike $(\mathrm{TMTSF})_2X$, the crystal structure of the ET salts can vary a lot. A few examples will be presented in the next section.

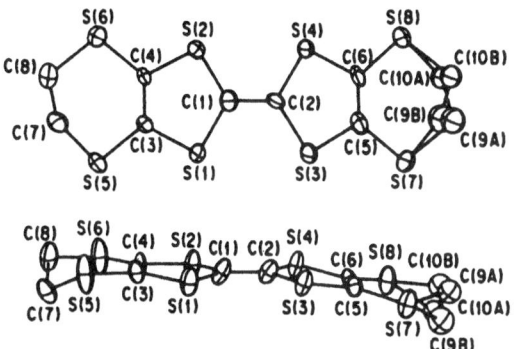

Fig. 2.13. ET molecule conformation taken in a charge transfer complex. Hydrogen atoms are not shown for clarity. The terminal ethylene groups C(9) and C(10) can have two different positions (A) and (B). From [125]

A large variety of other 2D superconducting salts with different organic constituents have been synthesized. Building blocks are for example DMET, MDT–TTF, BEDO–TTF or also the acceptor molecule $M(\mathrm{dmit})_2$ with $M =$ Ni, Pd, or Pt. These latter salts, however, will not be discussed in general, but rather when measurements of their FS will be presented.

In this section, after introducing some crystal structures, a selection of band structures calculated for different phases of the ET salts will be presented in Sect. 2.3.2. The band structures for some special materials will be presented in Chap. 4 together with the experimentally determined FS. In Sect. 2.3.3 an introduction to the superconducting properties of the 2D materials will be given.

2.3.1 Some Crystal Structures

In the charge transfer complexes $(\mathrm{ET})_m X_n$ the donor molecule ET shown in Fig. 2.13 is rather flat due to the existence of an extended π-electron system [125]. However, some deviations from planar geometry are present especially in the ethylene end groups. The different ways of "twisted" or "boat" conformations, their influence on the crystal packing, and the consequences on the physical properties have been discussed extensively [126].

A comprehensive overview of the large variety of crystal structures of the ET salts has been given in [33]. Therefore, only three main crystal structures of superconducting ET compounds will be shown here and discussed in more detail. Neither the crystal structures of other organic donors than ET nor

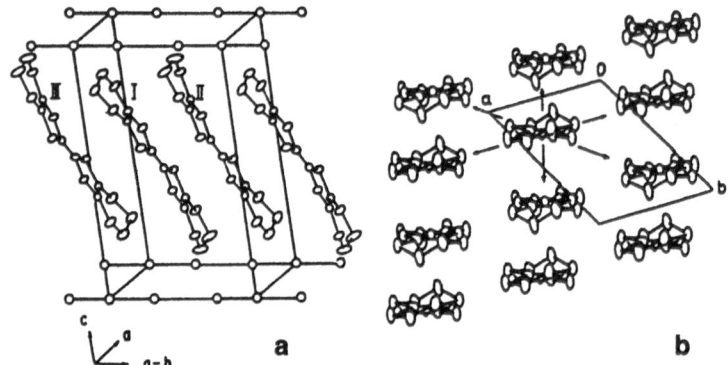

Fig. 2.14. Crystal structure of β-(ET)$_2$I$_3$ with (a) a perspective view showing the ET molecules sandwiched between the I$_3^-$ anions and (b) a view perpendicular to the ab plane showing the columnar stacking arrangement of the ET. From [128]

other acceptors will be shown here. The reader is referred to [28, 33] or the original literature for more recently synthesized materials.

As an example for the β-(ET)$_2X$ family the crystal structure of the salt with $X^- = $ I$_3^-$ is shown in Fig. 2.14 [127, 128]. β-(ET)$_2$I$_3$ was the second ambient-pressure organic superconductor ($T_c \approx 1.5$ K) to be found [129]. The principal structure of most ET salts can be seen in Fig. 2.14. The ET molecules are arranged in columnar stacks within planes which are separated by anion layers. The β-type structure is triclinic with the lattice parameters as given in Table 2.1. The ET molecules are stacked along the $a+b$ direction and are slightly dimerized. S–S contacts less than the van der Waals radii of 3.6 Å are found in the ab plane only between the stacks and not within the stacks. However, due to their 2D connectivity (see Fig. 2.14b) the ET form sheets of high conductivity in the ab planes which are separated by insulating layers of linear I$_3^-$ anions.

Other ET compounds with linear anions like IBr$_2^-$, AuI$_2^-$, and I$_2$Br$^-$ crystallize into the β structure. All of these are superconductors except β-(ET)$_2$I$_2$Br. The reason for this is thought to be a disorder of the asymmetrical anion I$_2$Br$^-$ in the crystal structure [131].

β-(ET)$_2$I$_3$ is also special in the β family in the sense that this salt undergoes a crystallographic phase transition at 175 K to an incommensurate modulated structure [132, 133]. The anion–cation interaction via hydrogen bonds of the terminal ethylene groups to the iodide is held responsible for the displacements [132]. By applying a moderate pressure of a few hundred bar it is possible to suppress the structural phase transition and to obtain the so-called β_H (or β^*) phase [134]. At the same time the superconducting transition temperature changes from $T_c \approx 1.5$ K for β-(ET)$_2$I$_3$ (also called β_L phase) to $T_c \approx 8$ K for β_H-(ET)$_2$I$_3$ [135].[5] At low temperatures the pressure

---

[5] Both cited T_c values are obtained from resistivity measurements. The actual thermodynamic critical temperature may differ. See also Sect. 4.2.2.

in the β_H phase can even be released with the sample staying in the "high-T_c" state as long as the temperature does not exceed $\sim 120\,K$ [136]. All other salts of the β family remain in the ordered state upon cooling.

Table 2.1. Room temperature crystallographic data and T_c for some two-dimensional organic metals based on the donor molecule ET. All compounds are of the form *phase*-$(ET)_2X$

phase − X	a(Å)	b(Å)	c(Å)	α(°)	β(°)	γ(°)
β-I$_3$	6.615	9.100	15.286	94.38	95.59	109.78
β-IBr$_2$	6.593	8.975	15.093	93.79	94.97	110.54
Θ-I$_3^a$	10.076	33.853	4.994	90	90	90
- superstr.	9.928	10.076	34.220	90	98.39	90
κ-I$_3$	16.387	8.466	12.832	90	108.56	90
κ-Cu(NCS)$_2$	16.256	8.456	13.143	90	110.28	90
α-TlHg(SCN)$_4$	10.051	20.549	9.934	103.63	90.48	93.27
α-KHg(SCN)$_4$	10.082	20.565	9.933	103.70	90.91	93.06
α-NH$_4$Hg(SCN)$_4$	10.091	20.595	9.963	103.67	90.47	93.30
α-KHg(SeCN)$_4$	10.048	20.722	9.976	103.59	90.43	93.26
α-RbHg(SCN)$_4$	10.087	20.642	9.998	103.54	90.53	93.23
α-TlHg(SeCN)$_4$	10.105	20.793	10.043	·103.51	90.53	93.27

phase − X	sp-gr.	V(Å^3)	Z	Ref.	$T_c(K)^b$
β-I$_3$	$P\bar{1}$	855.9	1	[127]	1.4
β-IBr$_2$	$P\bar{1}$	828.7	1	[130]	2.8
Θ-I$_3$	$Pnma$	1693	2	[147]	3.6
- superstr.	$P2_1/c$	3386	4	[149]	3.6
κ-I$_3$	$P2_1/c$	1687.6	2	[163]	3.5
κ-Cu(NCS)$_2$	$P2_1$	1694.8	2	[138]	10.4
α-TlHg(SCN)$_4$	$P\bar{1}$	1990	2	[150]	–
α-KHg(SCN)$_4$	$P\bar{1}$	1997	2	[141]	–
α-NH$_4$Hg(SCN)$_4$	$P\bar{1}$	2008	2	[140]	1.1
α-KHg(SeCN)$_4$	$P\bar{1}$	2015	2	[144]	–
α-RbHg(SCN)$_4$	$P\bar{1}$	2020	2	[151]	–
α-TlHg(SeCN)$_4$	$P\bar{1}$	2048	2	[143]	–

^{a} in this "average" structure approximately 2% of I$_3^-$ are replaced by AuI$_2^-$. See also [33].
^{b} here the usually stated values are given independent from the experimental determination of T_c. The actual thermodymanic values may differ slightly.

When correlating T_c and structural parameters of the β phase salts it was found that the linear anion length and T_c are proportional to each other [126]. Therefore, the next step to increase T_c was to search for ET salts with larger anions. Indeed, this search was successful with the discovery of

κ-$(ET)_2Cu(NCS)_2$, a superconductor with a resistively determined T_c of approximately 10.4 K [137]. The structure of the ET arrangement, however, is completely different from the β structure and was labeled κ phase. Figure 2.15 shows the crystal structure of this salt [138].

In the κ phase the most prominent difference from the other crystal structures is the way the donor molecules are packed within the highly conducting layer (here the bc plane). The ET molecules form dimers where adjacent pairs are rotated by approximately 90° with respect to each other and ~ 45° with respect to the b and c axes (Fig. 2.15b). Between neighboring ET sheets the anions, here V-shaped $Cu(NCS)_2^-$, form a separating insulation layer. In κ-$(ET)_2Cu(NCS)_2$ the anions are weakly connected via Cu–S bonds forming polymer-like zig-zag chains. Due to this V shape of the oriented anions there is no center of symmetry leading to a $P2_1$ space group for κ-$(ET)_2Cu(NCS)_2$. Other κ-phase salts such as κ-$(ET)_2I_3$ have centrosymmetric anions and correspondingly their crystal structure belongs to the $P2_1/c$ space group [139].

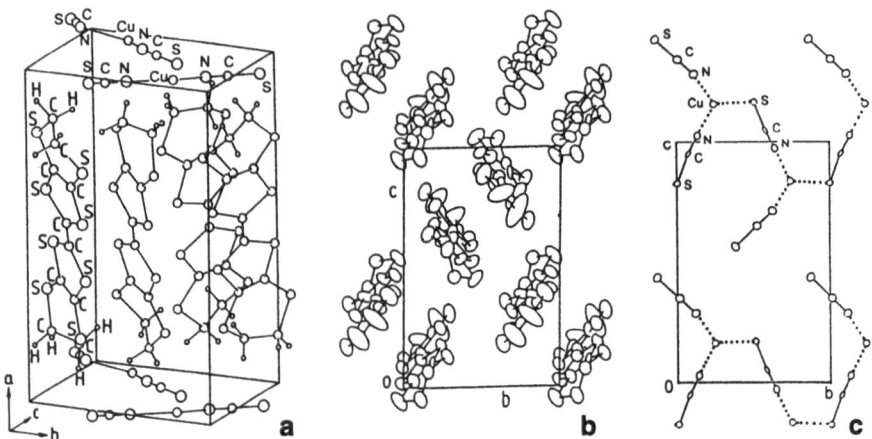

Fig. 2.15. (a) Crystal structure of κ-$(ET)_2Cu(NCS)_2$. (b) and (c) view along the a^* direction showing the arrangement of the ET molecules (b) and the $Cu(NCS)_2$ anions (c). From [138]

In an effort to increase both the anion size and T_c further, ET salts based on the anion MHg(SCN)$_4$ with M = NH$_4$, K, Tl, and Rb [140, 141, 142] and MHg(SeCN)$_4$ with M = Tl and K [143, 144] have been synthesized. This family, however, crystallizes in the so-called α phase which was known already from α-$(ET)_2I_3$ [145]. Except α-$(ET)_2NH_4Hg(SCN)_4$ with a T_c onset of ~ 1.15 K [146] none of these $(ET)_2M$Hg(YCN)$_4$ (Y = S or Se) salts becomes superconducting. However, due to their mixture of 1D and 2D bands and their unusual low-temperature ground states they have become the object of many different investigations.

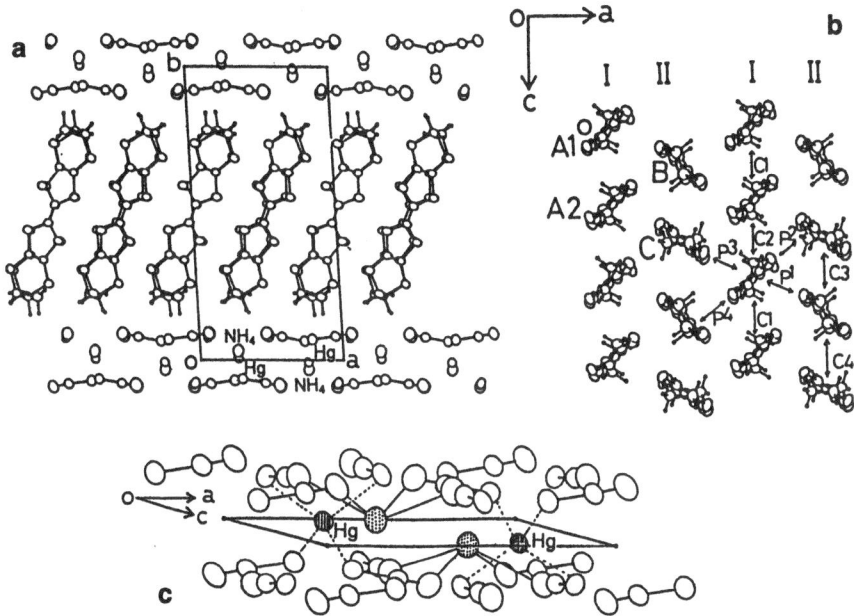

Fig. 2.16. (a) Crystal structure of α-(ET)$_2$NH$_4$Hg(SCN)$_4$, (b) view along the b^* axis (only the ET donors are shown). (c) Molecular arrangement of the anions. From [140]

Figure 2.16 shows the crystal structure of α-(ET)$_2$NH$_4$Hg(SCN)$_4$ [140]. The most obvious structural features are the "fish-bone" arrangement of the ET molecules in the conducting layer and the thick three-layered polymeric anion, two layers of thiocyanate and one with Hg^{2+} and NH$_4^+$ in between. The ET donors in the stack labeled II in Fig. 2.16b are located at nonequivalent inversion centers denoted by B and C, whereas the ET molecules in stack I are at equivalent positions ($A1$ and $A2$). Therefore, the unit cell contains two formula units. Within the α-phase family the unit-cell parameters are very similar with a slightly increasing cell volume from $M = $ Tl towards $M = $ Rb (see Table 2.1).

An analogous fish-bone structure of the ET donors similar to the α phase is found in the Θ phase of (ET)$_2$I$_3$ [147]. In this compound the I$_3^-$ ions are partly replaced by AuI$_2^-$ with incurring superconductivity at $T_c \approx 3.6$ K [148] (see also the discussion in [33]). There, however, two neighboring ET layers are nonequivalent. The molecules are tilted by $\sim 20°$ from plane to plane. Within the layer the donors in each stack are centrosymmetric leading to an orthorhombic structure. In a subsequent report, however, a monoclinic cell twice as large was proposed to describe all atomic positions of the iodine [149].

2.3.2 Band Structure

The general remarks in Sect. 2.2.2 given for the 1D organic metals are in principle also relevant for the 2D materials. In TMTSF, however, the inter-molecular electronic overlap is due to the π electrons and hence always in the stacking direction resulting in the quasi-1D band. From the structures shown in the previous section it is obvious that for the ET salts the overlap must be drastically different from one phase to the other. Whether the ET molecules are aligned face-to-face or side-by-side changes the overlap integral and consequently the transfer energies considerably. The overlap integrals of the HOMO bands between two neighboring TTF or ET molecules for fixed distance have been calculated as a function of angle in general by T. Mori [152]. From the overlap integral it is possible to obtain the transfer energy, t. With the appropiate values of t, knowledge of the crystal structure (which may change at low temperatures), and the usual tight-binding approximation, most of the following band structures have been obtained.

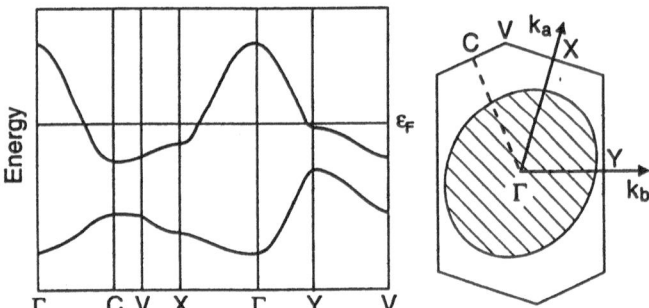

Fig. 2.17. Calculated band dispersion relation for the two highest occupied bands and the FS of β-(ET)$_2$IBr$_2$. From [153]

Figure 2.17 shows the calculated energy dispersion relation and the FS of β-(ET)$_2$IBr$_2$ [153, 154]. The HOMO is half filled and due to the isotropic overlap integrals between the ET molecules within the 2D layer (indicated by the arrows in Fig. 2.14 for the isomorphous β-(ET)$_2$I$_3$) the resulting FS is almost circular as expected from a free-electron picture. Calculations by other groups using a finer mesh of k points have yielded a slightly higher asymmetry of the FS but otherwise only gradual changes of the dispersion relation, the shape, and extremal cross-sectional area of the FS [155, 156].

As mentioned above, the β phase of (ET)$_2$I$_3$ undergoes a structural order–disorder transition at 175 K. Based on the different crystal structures at 9 K under ambient pressure [157] and at 4.5 K under 1.5 kbar [158], the band structures have been determined [155]. However, virtually no difference is found from the calculations.

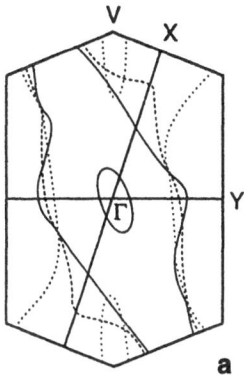 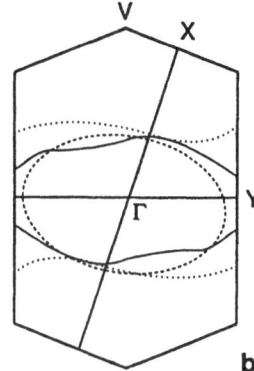

Fig. 2.18. Early (a) and later refined (b) calculation of the Fermi surface of β-$(ET)_2IBr_2$ based on the local density approximation. The different lines represent intersections of the FS with the planes $k_c = 0$ (solid lines), $k_c = 0.25$ (dashed lines), and $k_c = 0.5$ (dotted lines). From [159] and [160]

From the beginning of band-structure calculations of organic metals the applicability of the extended Hückel tight-binding method was questioned. This approximation is powerful mainly when the bandwidth of the resulting molecular levels is less than their separation. Therefore, attempts were made to calculate the band structure *ab initio* based on the local density functional approximation (LDA). The main results of a first preliminary [159] and a later refined calculation [160] are shown in Fig. 2.18. Since in this treatment the out-of-plane interactions have been taken into account as well a true 3D energy dispersion and FS have been obtained. In the first approximation shown in Fig. 2.18a obviously too many simplifications have been employed. The FS is far from being the inherently perfect 2D cylinder of Fig. 2.17. Small closed portions and even open sheets coexist leading to a more pronounced 3D topology. The improved later calculation shown in Fig. 2.18b resulted in a band structure closer to the one obtained by the extended Hückel approximation but still with appreciable k_c dispersion. As will be discussed in more detail in Sect. 4.2, the experimentally determined FS is often in very good agreement with the latter crude tight-binding calculation. In principle, however, the LDA should yield more reliable results. Yet, the large number of atoms per unit cell (55) in a complicated crystal structure, the low symmetry $(P\bar{1})$, the low density with a lot "empty" space (the atomic spheres used in the LDA treatment occupy only roughly one third of the total volume), and limited computer power all force some simplifications which may lead to an incorrect FS topology.

Based on the extended Hückel tight-binding method the 2D energy dispersion relation and FS of κ-$(ET)_2Cu(NCS)_2$ [29, 155, 161, 162] and κ-$(ET)_2I_3$ [147, 163] have also been calculated. Figure 2.19 shows the results. The band structures are very similar except for the degeneracy of the two upper bands along Z–M for κ-$(ET)_2I_3$. In κ-$(ET)_2Cu(NCS)_2$, due to the lack of a center of

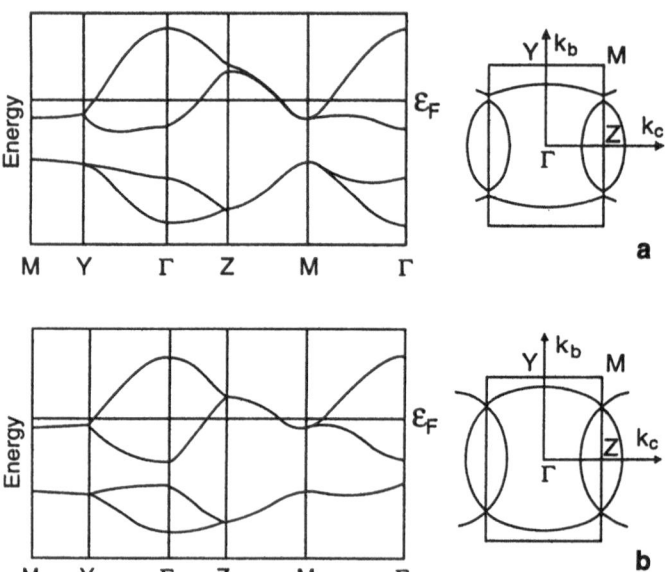

Fig. 2.19. Calculated energy dispersion and Fermi surface of (a) κ-(ET)$_2$Cu(NCS)$_2$ [29, 155, 161, 162] and (b) κ-(ET)$_2$I$_3$ [147, 163]

symmetry, these bands split up and a small gap opens at the zone boundary (exaggerated in Fig. 2.19a). Of course, also in κ-(ET)$_2$I$_3$ small gaps due to Bragg reflections at the zone boundary exist. This gap, however, is considerably smaller than in the former salt, although it is observable in dHvA and SdH experiments (see below).

Due to the gap opening the FS consists of a closed hole-like orbit and two open corrugated sheets [164], giving rise to an observable in-plane anisotropy of these quasi-2D materials. Experimentally, this anisotropy could easily be detected, e. g., by optical reflectivity measurements [165].

It should be noted here that the overall shape of the κ-phase FS can also be obtained by a simple free-electron treatment. With the usual parabolic bands and the known electron density one obtains a circular FS which cuts the Brillouin zone at approximately the point where the calculated gaps in Fig. 2.19 occur. Folding back these FS parts into the first Brillouin zone results in an only slightly modified topology compared to the calculated tight-binding FS of κ-(ET)$_2$I$_3$. The effective masses estimated from the predicted band-structures are close to the free-electron mass, m_e. These values, however, are in contradiction to the experimentally extracted masses from optical [165, 166] and also dHvA or SdH measurements (see Sect. 4.2).

As a last example for a calculated band structure in this chapter, Fig. 2.20 shows the result for an α-phase salt, namely α-(ET)$_2$KHg(SCN)$_4$, based on the crystal structure at $\sim 100\,\mathrm{K}$ [141]. The other members of the α-(ET)$_2$$M$Hg(SCN)$_4$ family are expected to have almost the same FS. Only

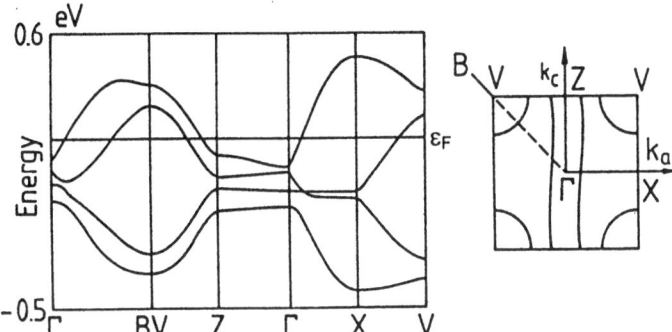

Fig. 2.20. Calculated band structure and Fermi surface of α-(ET)$_2$KHg(SCN)$_4$. From [141]

the extremal cross-sectional area of the FS might be slightly changed due to the different unit cell volumes. Similar to the κ phase, 1D and 2D bands coexist, two lightly warped open sheets with the surface vector running approximately along the a direction and a closed hole-like orbit in the corner

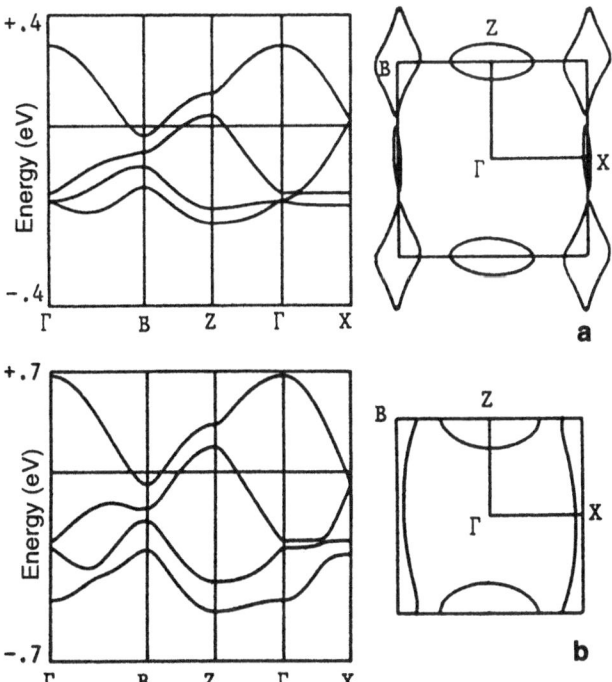

Fig. 2.21. single-ζ (a) and double-ζ (b) band structures and Fermi surfaces of α-(ET)$_2$KHg(SCN)$_4$. From [169]

of the Brillouin zone. However, in contrast to the κ phase but comparable to the $(TM)_2X$ salts the 1D open sheets in α-$(ET)_2MHg(SCN)_4$ exhibit almost perfect nesting. Indeed, the salts with M = K, Rb, and Tl are unstable against an electronic phase transition around 8–12 K as first evidenced by a resistivity hump at this temperature [142, 167]. Early suggestions of a SDW phase transition with a consequently reconstructed FS seem to be verified by recent μSR measurements [168]. Probably due to this reconstruction of the FS, however, the results and interpretations of dHvA and SdH data are still very controversial (see Sect. 4.2.1).

Recently, the correctness of the band structure of the α phase shown in Fig. 2.20 was called into question. A careful analysis of the sensitivity of the band structure on the approximations of the transfer integrals, t_i, revealed, especially for the α phase, drastic changes of the FS [169]. Figure 2.21 shows the band structure and FS of α-$(ET)_2KHg(SCN)_4$ for the so-called single-ζ (a) and double-ζ (b) approximation. As is evident from this figure, the slight modifications in t_i change the FS topology completely. Whether this is the reason for the somewhat inconsistent experimental results is unclear and more systematic studies are necessary to resolve this question.

For many other organic metals the band structures and FS topologies have been calculated by tight-binding methods. Some of these results will be presented and discussed in Chap. 4 in connection with and comparison to the experimental data.

Except for the mentioned *ab initio* band-structure calculations by Kübler et al. [159, 160] the overlap and transfer integrals, $t_\perp$, perpendicular to the highly conducting planes are neglected. Therefore, the band structures shown are 2D bands with a cylindrical FS. However, a remaining overlap and a weak dispersion in the third spatial direction (not always c, depending on the crystal structure) should exist in principle. Indeed, in a variety of different experiments, the influence of the very weak transfer integrals could be detected. The usual description of the energy dispersion can be given by

$$\epsilon_k = \frac{\hbar^2}{2m_b}(k_x^2 + k_y^2) - 2t_\perp \cos(\boldsymbol{hk}), \qquad (2.10)$$

where for simplicity we have assumed a circular free-electron in-plane FS with an effective mass m_b, $\boldsymbol{h} = (u_x, u_y, c')$ is the direction vector of the interlayer transfer energy $t_\perp$, and c' is the spacing between adjacent layers. Simplifying further, we assume that $\boldsymbol{h}$ has no in-plane components, that is u_x and u_y are zero, being aware that in crystal structures of lower symmetry an oblique $\boldsymbol{h}$ can be found (see Sect. 4.2).

A schematic view of this corrugated FS cylinder is shown in Fig. 2.22a. The oscillation frequencies observed in the dHvA (or SdH) effect are proportional to the extremal cross-sections of the FS (see also Chap. 3). Therefore, when measuring a metal with a warped FS as shown in Fig. 2.22a two slightly different frequencies with a resultant beating behavior will be observed. Depending on the direction of the applied field B, however, the frequency difference will

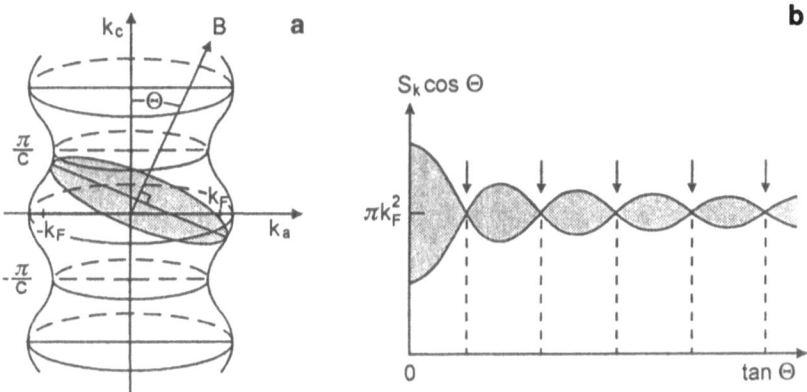

Fig. 2.22. (a) Schematic view of a corrugated 2D Fermi surface. (b) Angular dependences of the extremal cross-sections, S_k, according to (2.11) plotted against $\tan\Theta$. From [170]

change and eventually become zero. As was first calculated analytically by Yamaji, the angular dependence of this frequency difference ΔF for $t \ll \epsilon_F$ can be written as [170]

$$\Delta F \cos\Theta = \Delta F_0 J_0(c' k_F \tan\Theta),   (2.11)$$

where $\Delta F_0 = 4 m_b t_\perp / e\hbar$ is the maximum frequency difference, J_0 is the Bessel function of zeroth order and k_F is the planar Fermi wave vector. The principal behavior of the angular dependence of the extremal cross-sectional FS areas according to 2.11 is shown in Fig. 2.22b. With the approximation for large arguments of the Bessel function the angles where $\Delta F = 0$ are given by

$$\tan\Theta_n = \frac{\pi}{c' k_F}\left(n - \frac{1}{4}\right),   (2.12)$$

where n is an integer. At these special angles the whole FS has only one extremal cross-sectional area. Therefore, at these topological points unusual effects are observable. First, the amplitude of the magnetic quantum oscillations becomes extremely large because all electrons in the highest occupied Landau level are contributing to the oscillatory effect. Second, when measuring the resistance, R, in the z direction while rotating the sample in an applied magnetic field a new kind of oscillatory behavior of $R(\Theta)$ periodic in $\tan\Theta$ can be observed. At the angles Θ_n the electron velocity component in the z direction becomes zero resulting in a large increase of R at these points [170, 171]. In Sect. 3.3 it will be described how this technique can be used to map out the FS in-plane topology [172].

2.3.3 The Superconducting State

The 2D ET salts comprise the largest number of organic superconductors. Consequently an enormous number of investigations on the superconducting properties exist. However, the nature of the superconducting state in these organic materials is far from being understood. The question whether it is of the usual BCS type, possibly with strong coupling, or of more exotic origin is not yet clear. This unsatisfactory situation is partly caused by the lack of consistent or definitive experimental results. The small crystal sizes usually available, the dependence of the superconducting properties on sample preparation and history, the strongly type-II superconducting behavior, and so on contribute to the serious experimental problems involved in measuring and extracting the superconducting properties. This can already be seen in the sometimes quite different reports of the superconducting transition temperatures. Besides the changes of T_c from sample to sample different measuring methods reveal different values. Especially, the T_cs determined in magnetic fields by resistivity measurements are highly ambiguous. This effect is known from the cuprate superconductors and T_cs given just by the onset, the midpoint, or zero value may have little to do with the true thermodynamic T_c. Therefore, derived quantities like critical fields, coherence lengths, etc are only approximate. Furthermore, thermodynamically consistent determinations of these important characteristics of superconductors are missing for many of the organic materials.

In this section some of the experimental controversies but also the principal superconducting properties generally agreed upon will be reviewed. Of course, not all aspects of the diversity of existing experimental data can be presented. In the following the results will be organized by the investigated property rather than by the different phases or systems.

The strongly 2D nature of the ET salts is reflected in many superconducting properties. The layered structure of the materials causes anisotropic coupling as seen by the directional dependence of the critical fields, the coherence length, the penetration depth, and so on. Torque measurements, for example, have suggested a lower bound of $\Gamma \approx 4 \times 10^4$ for the anisotropy parameter, i. e., the effective mass anisotropy [173]. It was pointed out, however, that the way the above mentioned value was determined might be invalid for strongly anisotropic superconductors [174]. Nevertheless, other thermodynamic experiments agree that Γ is of the order $10^3 - 10^4$.

Similar to TMTSF salts the superconductivity is very sensitive to external pressure. As mentioned already, one of the most sensitive materials is β-$(ET)_2I_3$ with the "high-T_c" state under moderate pressure of a few hundred bar. Figure 2.23 shows the pressure dependence of T_c for different β-phase salts obtained from resistivity measurements [175]. For β-$(ET)_2I_3$ at approximately 0.4 kbar the abrupt change in T_c from only just above 1 K to ~ 7 K is seen. The other main feature common to all ET superconductors is the unusually large decrease of T_c with pressure, $\Delta T_c/\Delta P \approx -1$ K/kbar, which is one order of magnitude larger than in $(TMTSF)_6PF_6$ [101]. This large pressure dependence of T_c is correlated with the anion length dependence of T_c. Using the compressibility data of β-$(ET)_2I_3$ obtained at room temperature by

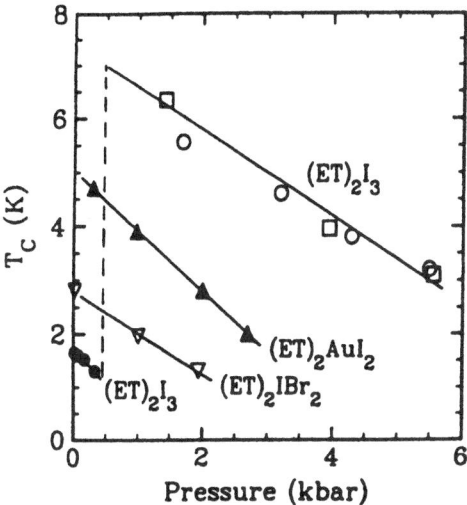

Fig. 2.23. Pressure dependence of T_c for different salts of the β phase. From [175]

X-ray diffraction [176], the different T_cs are explicable with so-called "lattice pressure" which means the larger the unit cell volume due to the larger anion length the larger is T_c. This is nicely seen by the nearly linear relationship between anion length and T_c for these β salts as shown in Fig. 2.24. Therefore, the "high-T_c" state of β-$(ET)_2I_3$ (I_3^- has the largest anion length) is believed to be the "true" T_c of this compound which is suppressed by the incommensurate lattice modulation [126]. For β-$(ET)_2I_2Br$, which also has a large anion length, disorder of the anions completely destroys superconductivity.

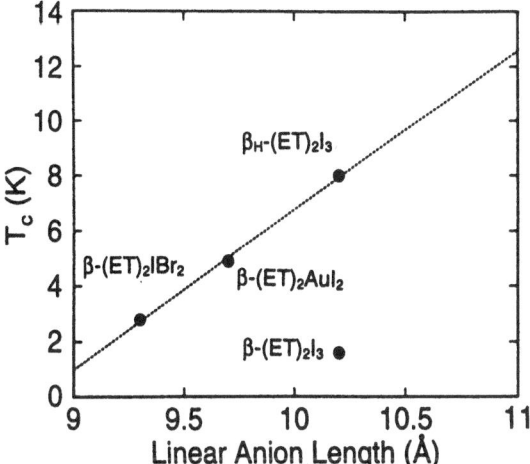

Fig. 2.24. T_c dependence in the β-type $(ET)_2X$ salts with linear, triatomic anions vs anion length (sum of bond length and twice the van der Waals radius of the terminal anion). From [126]

The largest pressure dependence yet observed for any superconductor is found in κ-(ET)$_2$Cu(NCS)$_2$ with $\Delta T_c/\Delta P \approx -3$ K/kbar [177]. From thermal expansion measurements the uniaxial pressure dependence was derived which showed the extreme sensitivity of T_c under pressure along the a^* direction, i.e., the direction perpendicular to the highly conducting planes. For κ-(ET)$_2$Cu(NCS)$_2$ a value of $\partial T_c/\partial p_{a^*} \approx -4.8$ K/kbar [178] and for κ-(ET)$_2$Cu[N(CN)$_2$]Br ($T_c \approx 11$ K) $\partial T_c/\partial p_{a^*} \approx -2.4$ K/kbar [179], the same value as observed under hydrostatic pressure [180] was found. These results seem to indicate that as long as one stays within the same crystal phase the enlargement of the anion layer thickness might increase T_c even further.

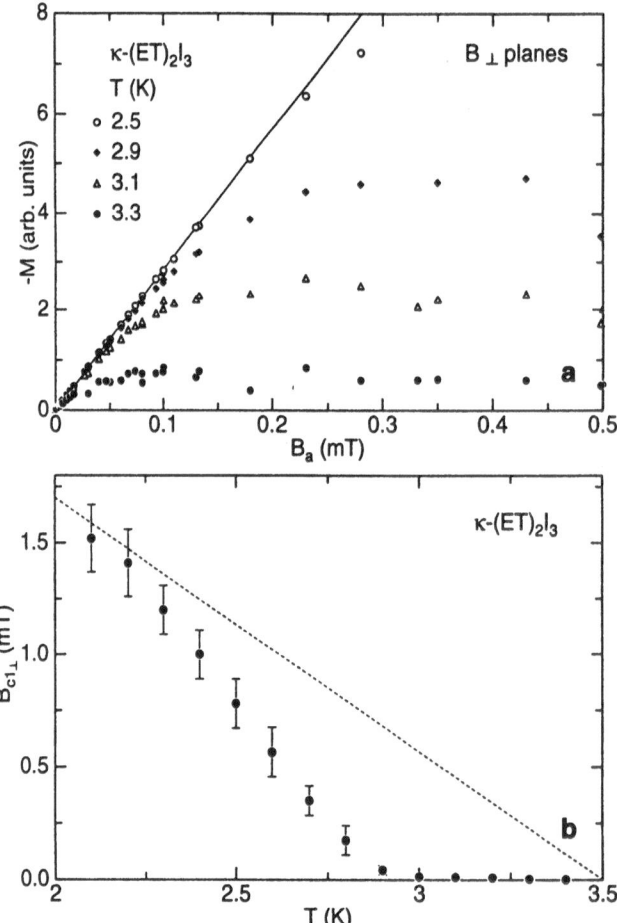

Fig. 2.25. (a) Magnetization of κ-(ET)$_2$I$_3$ vs magnetic field applied perpendicular to the ET planes for different temperatures. **(b)** Temperature dependence of B_{c1} obtained from the points where the magnetization curves **(a)** first deviate from ideal diamagnetism (solid line in **(a)**)

Another characteristic feature of the 2D organic superconductors is their large Ginzburg–Landau (GL) parameter κ, that is their very small lower critical field B_{c1} and an upper critical field B_{c2} often close (or even above) the Pauli limit (see Sect. 2.2.4). The usual and most reliable way to determine B_{c1} is by measuring the field dependent magnetization. This has been done for a number of compounds partly to establish the volume superconductivity in those salts [181, 182]. As an example, Fig. 2.25a shows the magnetization of κ-$(ET)_2I_3$ ($T_c = 3.5\,K$) against the applied magnetic field B parallel ($\parallel$) a^*, i. e., perpendicular ($\perp$) to the ET planes, for different temperatures [183]. The data were measured in a home-built low-field SQUID magnetometer and are obtained from temperature sweeps at a constant field. The sample was first cooled in zero field (the Earth's field was shielded by a μ-metal cylinder and the residual field compensated by applying a small counter-field), then at low temperatures the desired field was applied (shielding mode). The solid line in Fig. 2.25a represents the perfect diamagnetic behavior calculated from the known sensitivity of the apparatus and the estimated demagnetization factor. Already at very low field the magnetization curves deviate from this line and after passing over a broad maximum tend to decrease again. For the temperatures shown B_{c2} was not reached in this experiment. Meissner measurements (cooling and heating in field) gave for the lowest fields approximately 70% of the shielding signal. This shows the very good quality of the sample.

$B_{c1}(T)$ defined as the first deviation of each measured curve from perfect diamagnetism was extracted by plotting the difference between data and the solid line in Fig. 2.25 [183]. The somewhat surprising result is shown in Fig. 2.25b. From $T_c = 3.5\,K$ down to $\sim 3\,K$ the critical field is almost zero. Then it rises approximately linearly with a slope of $\Delta B_{c1\perp}/\Delta T \approx -1.5\,mT/K$. A similar B_{c1} behavior has been observed for the strongly layered cuprate superconductor $Bi_2Sr_2CaCu_2O_8$ [184]. In a recent theory this was explained as a characteristic feature of strongly 2D superconducting materials. In an applied field order-parameter fluctuations within single vortices are assumed to be the origin of the reduced T_c [185]. Although a tendency for this upward curvature of B_{c1} towards lower temperatures has also been found in κ-$(ET)_2Cu(NCS)_2$ [182] this sharp and distinct feature shown in Fig. 2.25 is the first such kind found in organic superconductors.

From the B_{c1} curve for $B \parallel a^*$ one can try to estimate a value for $B_{c1\perp}$ at $T = 0$ which yields approximately $(4 \pm 1)\,mT$. This extrapolated field is consistent with $B_{c2\perp}$ values and specific-heat results as will be shown below. For fields applied within the ET planes of κ-$(ET)_2I_3$ ($B \perp a^*$) $B_{c1\parallel}$ starts to grow at T_c. In accordance with the usually observed behavior for 3D materials B_{c1} increases nearly linearly with a slope $\Delta B_{c1\parallel}/\Delta T \approx -0.01\,mT/K$ observed down to $2\,K$ [183].

The upper critical field, B_{c2}, for ET materials was determined by many different groups and in different ways. Most results are obtained by resistivity measurements. However, in magnetic fields the effect of fluctuations in the resistivity becomes increasingly important, especially for layered superconductors. This has been studied for classical layered materials [186] and was even more evident from transport and thermodynamic measurements of cuprate high-T_c material [187]. For these latter compounds a scaling analysis based on calculations of Ref. [188] using the Lawrence–Doniach model [107] was performed. This method has recently also been used to extract $B_{c2}(T)$ for κ-(ET)$_2$X salts [189].

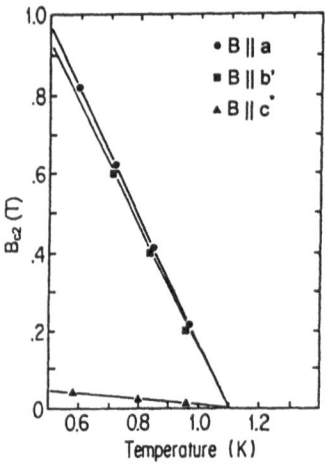

Fig. 2.26. Temperature dependence of the upper critical field of β-(ET)$_2$I$_3$ along three directions determined by resistivity measurements. From [190]

Nevertheless, the anisotropy of B_{c2} can be inferred from the transport data. Figure 2.26 shows early results for the low-T_c state of β-(ET)$_2$I$_3$ [190]. $B_{c2}(T)$ defined as the midpoint of the resistive transitions was determined for the three principal crystallographic axes. At $\sim 0.5\,\mathrm{K}$ the critical fields within the ET planes (ab plane) are approximately 20 times larger than B_{c2} along the c^* axis. Determinations of B_{c2} with a tunnel diode oscillator technique yielded considerably higher values in all directions [191] showing the ambiguity in the extraction of the critical field by different measuring methods.

A slight anisotropy of B_{c2} for fields within the ab plane is seen in both experiments. Although the sample has been aligned with great care a small misalignment of only 1°–2° cannot be excluded. This, however, already has a drastic influence on B_{c2} which can be seen in Fig. 2.27 [183, 192], where the angular dependence of the critical field obtained from ac-susceptibility measurements at two temperatures is plotted. In distinction to the thermodynamic value B_{c2} this field is labeled B_{c2}^*. For $T = 0.55\,\mathrm{K}$ it was possible to measure the critical fields very close to 90° ($B \perp a^*$) with high angular

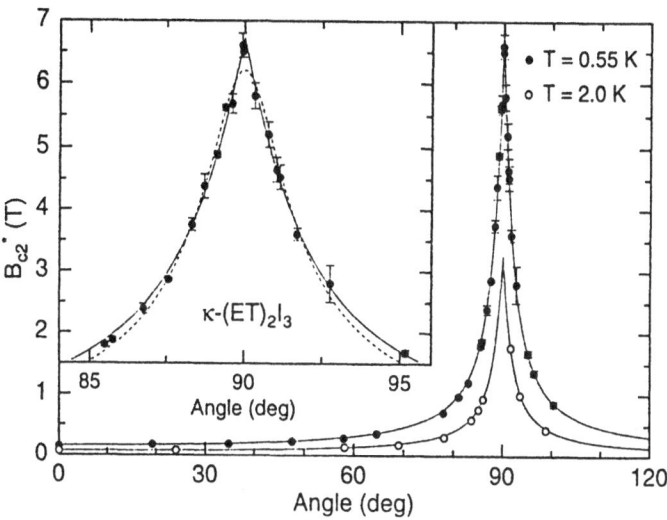

Fig. 2.27. Angular dependence of the critical field extracted from ac-susceptibility measurements of κ-(ET)$_2$I$_3$ for two temperatures. The inset shows the region close to $\Theta = 90°$ (i.e., B within the plane) on an enlarged scale. The solid line is a fit with the 2D Tinkham formula (2.13), the dashed line with the anisotropic mass model (2.14)

resolution. The critical fields at this temperature for $\Theta = 0°$ and $\Theta = 90°$ are $B_{c2\perp} = 0.15$ T and $B_{c2\parallel} \approx 6.7$ T, respectively. The solid lines in Fig. 2.27 represent a fit derived for 2D films with $\xi \gg d$ (d = layer thickness) and are given by the implicit relation [193, 194]

$$1 = \left| \frac{B_{c2}(\Theta) \cos \Theta}{B_{c2\perp}} \right| + \left(\frac{B_{c2}(\Theta) \sin \Theta}{B_{c2\parallel}} \right)^2 . \tag{2.13}$$

This relation should also be valid for $\xi \ll c'$, where c' is the interlayer thickness. The perfect fit of this 2D model to the data is obvious. The dashed line in the inset is calculated using the anisotropic GL theory ($\xi > c'$) which relates the angular dependence of B_{c2} to strongly anisotropic effective masses by [105]

$$B_{c2}(\Theta) = \frac{\phi_0}{2\pi\xi_\parallel^2 \sqrt{\cos^2 \Theta + \frac{m_\parallel}{m_\perp} \sin^2 \Theta}} , \tag{2.14}$$

where $m_\parallel$ and $m_\perp$ are the GL effective masses within the ET planes and perpendicular to the planes, respectively. They are connected to the critical fields by

$$\frac{B_{c2\perp}}{B_{c2\parallel}} = \left(\frac{m_\parallel}{m_\perp} \right)^{\frac{1}{2}} . \tag{2.15}$$

This model, however, clearly deviates from the experimental data. This is decisive evidence for a perfectly 2D superconducting behavior in the ET compounds. Previous experiments could not definitely distinguish between the two models [191, 195].

For strong type-II superconductors a difference may exist between B_{c2}^* obtained from ac-susceptibility data and the thermodynamically relevant upper critical field, B_{c2}, for which (2.13) and (2.14) were derived originally. From extensive work on high-T_c cuprates it is known that B_{c2}^* is related to the so-called irreversibility line [196]. The basic idea of a semiquantitative flux-creep theory [197] is that pinned vortices can be activated thermally over an energy barrier U_0 resulting in a reduced critical current of the form [196]

$$J_c = J_{c0}[1 - (k_B T/U_0)\ln(f_0/f)], \qquad (2.16)$$

where J_{c0} is the critical current in the absence of thermal activation, f is the measuring frequency, and f_0 is a characteristic attempt frequency. The irreversibility line $B_{c2}^*(T, f)$ is defined by $J_c = 0$. U_0 scales as the condensation energy for a characteristic volume $B_{cth}^2 V/2\mu_0$, where μ_0 is the permeability of vacuum. Usually the thermodynamic critical field B_{cth} near T_c scales like $B_{cth} \propto (1 - t)$, with $t = T/T_c$. Different models exist to estimate the characteristic excitation colume V [196]. One possible assumption is that $V \propto a_0^3$, where $a_0 = 1.075\sqrt{\phi_0/B}$ is the flux-line spacing in a field B. Therefore, $U_0 \propto (1 - t)^2/B^{3/2}$. Approximating T by T_c and solving (2.16) for $J_c = 0$ results in

$$B_{c2}^* \propto (1 - t)^{4/3}[\ln(f_0/f)]^{-2/3}. \qquad (2.17)$$

This equation describes the characteristic temperature and frequency dependence of B_{c2}^* found for $\kappa\text{-(ET)}_2 I_3$.

Figure 2.28 shows B_{c2}^* vs modulation-field frequency for two different field orientations at $T = 0.55\,\text{K}$ [183]. The dashed lines are fits using (2.17) for fixed temperature. For the characteristic attempt frequency f_0 extraordinary large values of $f_0 \simeq 10^{17}\,\text{Hz}$ and $f_0 \simeq 10^{19}\,\text{Hz}$ were found for the data at $\Theta = 0°$ and $\Theta = 83°$, respectively. For temperatures closer to T_c the relative changes of B_{c2}^* with frequency become larger and the fitted values of f_0 decrease continuously towards $f_0 \approx 5 \cdot 10^6\,\text{Hz}$ at $T = 3.29\,\text{K}$. For the temperatures shown in Fig. 2.28 the relative increase of B_{c2}^* with f is the same for both angles within error bars. This indicates that the angular dependence of B_{c2}^* shown in Fig. 2.27 is likely to be the same as that of B_{c2} normalized by an appropriate factor.

The temperature dependence of B_{c2}^* is shown in Fig. 2.29 by open circles [183]. The data reveal a positive curvature which has been observed for many other organic superconductors by ac-susceptibility and resistivity measurements. The solid line is a one-parameter fit using (2.17) for fixed f which describes the data down to the lowest temperature. The exact temperature dependence of B_{c2}^* depends on the detailed nature of the pinning centers which

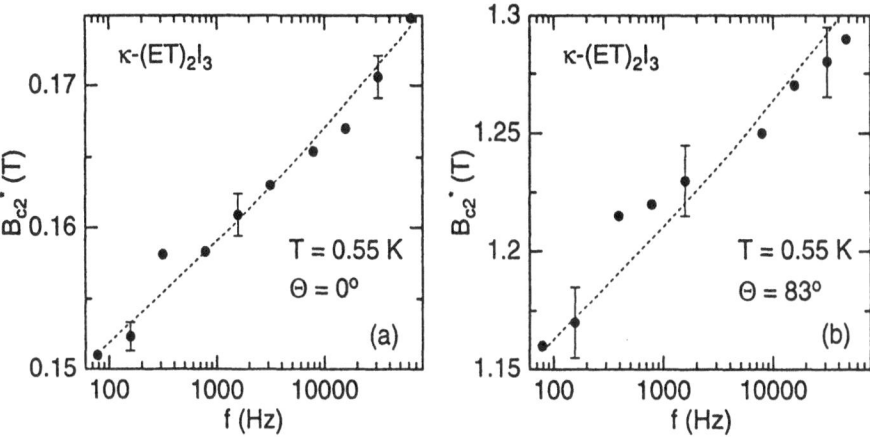

Fig. 2.28. Frequency dependence of the critical field of κ-(ET)$_2$I$_3$ extracted from ac-susceptibility measurements at $T = 0.55\,\mathrm{K}$ for (a) $\Theta = 0°$ and (b) $\Theta = 83°$. The dashed lines represent fits of the form $B_{c2}^* \propto \ln(f_0/f)^{-2/3}$

may vary between different materials and from sample to sample. From magnetization measurements for κ-(ET)$_2$Cu(NCS)$_2$ and κ-(ET)$_2$Cu[N(CN)$_2$]Br an exponential dependence, $B_{c2}^* \propto \exp(-AT/T_c)$, with fitting parameter A was found [189]. These results show clearly that for organic superconductors similar to the high-T_c cuprates the critical fields extracted from ac-susceptibility measurements are different from the thermodynamic upper critical field. B_{c2}^* rather describes the field where flux lines get pinned and are no longer able to follow field variations within the time scale set by the measuring frequency.

Figure 2.29 also shows the critical fields determined by magnetization measurements [183] which concide nicely with specific-heat data (see below) [198]. For the temperature range between $2.4\,\mathrm{K} < T < 3.25\,\mathrm{K}$ a roughly linear T dependence with a slope of $\sim 0.061\,\mathrm{T/K}$ is found. In this temperature range B_{c2} is clearly larger than the B_{c2}^* values extracted from ac susceptibility. Close to T_c an unexpected behavior is observed. Above $3.25\,\mathrm{K}$ the temperature dependence of B_{c2} becomes much steeper up to $\sim 3.38\,\mathrm{K}$ where a shallow tail can be seen (inset of Fig. 2.29). Similar features have been reported for κ-(ET)$_2$Cu[N(CN)$_2$]Br [199] and for YBa$_2$Cu$_3$O$_{7-\delta}$ [200]. In the former experiment a dimensional crossover from strong anisotropy ($B_{c2\parallel}/B_{c2\perp} \equiv \gamma_a \approx 80$) at low temperature to a weaker one ($\gamma_a \approx 13$) close to T_c was suggested. For κ-(ET)$_2$I$_3$ from T_c down to $\sim 3.3\,\mathrm{K}$ the $B_{c2\parallel}$ data (open triangles in the inset of Fig. 2.29) are even below $B_{c2\perp}$, suggesting a reversed anisotropy. At lower temperatures $B_{c2\parallel}$ grows much faster than $B_{c2\perp}$, reaching approximately $7\,\mathrm{T}$ ($\gamma_a = 35$) at the lowest temperature. This behavior might be induced by a dimensional crossover but more likely it is related to a rapid initial increase of fluctuations seen in the magnetization measurements which cannot be

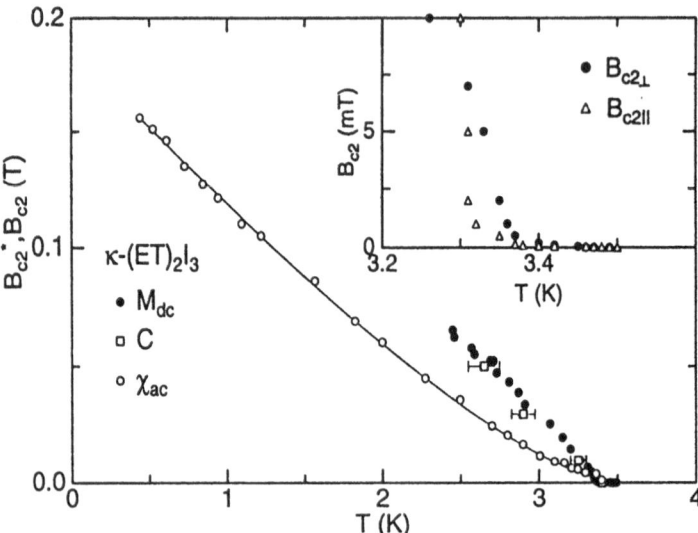

Fig. 2.29. Temperature dependence of the critical fields of κ-(ET)$_2$I$_3$ obtained from magnetization, specific heat, and ac-susceptibility data. The solid line is a fit to the ac-susceptibility data according to (2.17). The inset shows the temperature dependence of B_{c2} obtained from magnetization data close to T_c in an enlarged scale

taken into account properly [183]. Already in very small external fields (less than Earth's field) the onset of the diamagnetic magnetization signal shifts towards higher temperatures and the fluctuation-induced rounded region of the $M(T)$ curves increases considerably. A detailed theoretical understanding of the temperature-dependent magnetization for layered superconductors in this low-field range is missing.

The 2D behavior of $B_{c2}(\Theta)$ shown in Fig. 2.27 implies that in κ-(ET)$_2$I$_3$ the superconducting ET layers are decoupled from each other at least at low temperatures. This can also be seen from the coherence lengths $\xi_\parallel$ and $\xi_\perp$. With the extrapolated values of B_{c2} for $T \rightarrow 0$ and the anisotropic GL model these quantities can be calculated by rewriting (2.7) (ignoring the weak in-plane anisotropy):

$$\xi_\parallel = \sqrt{\frac{\phi_0}{2\pi B_{c2\perp}}} \tag{2.18}$$

$$\xi_\perp = \frac{\phi_0}{2\pi B_{c2\parallel}\xi_\parallel}. \tag{2.19}$$

The resulting values are given in Table 2.2. The coherence length $\xi_\perp$ between the layers is 1.1 nm which is smaller than the interlayer spacing of ~ 1.55 nm. In a previous work from resistivity data similar values ($\xi_\parallel = 36$ nm and $\xi_\perp = 1.35$ nm) have been obtained [201].

Another interesting point is the comparison of $B_{c2\parallel}$ with the Pauli limiting value of $B_P \approx 6.5$ T for $T_c = 3.5$ K. The experimentally obtained value is slightly higher than this limit. However, as was pointed out in the original work [108] the actual value of B_P is increased by the factor $(1 + \lambda)^{1/2}$, where λ is the electron–phonon coupling constant [202]. Other experiments that will be discussed later suggest a tendency towards strong coupling for the ET superconductors. Therefore, the real value for B_P might be appreciably larger than the one estimated from the simple expression (2.8). Thus, the large $B_{c2\parallel}$ is still in accord with standard s-wave pairing.

Table 2.2. Superconducting parameters of different κ-(ET)$_2 X$ compounds

$X =$	I_3	$Cu(NCS)_2$	$Cu[N(CN)_2]Br$
T_c (K)	3.5	9.1	11.0
$B_{c2\perp}$ (T)	0.2	6[a]	8[a]
$B_{c2\parallel}$ (T)	7	60[b]	44[b]
$B_{c1\perp}$ (mT)	4	6.5	1.6
$B_{c1\parallel}$ (mT)	0.05	0.2	
B_{cth}^c (mT)	20	100	50
B_{cth}^d (mT)	17	90[e]	
$\kappa_\perp$	7	42	110
$\kappa_\parallel$	675	1020	
$\xi_\perp$ (nm)	1.1	0.7	1.2
$\xi_\parallel$ (nm)	41	7.4	6.4
$\lambda_\perp$ (nm)	300	535[f]	650[g]
$\lambda_\parallel$ (μm)	70	22	
Ref.	[183]	[189, 203]	[189, 203]

[a]crudely extrapolated values.
[b]with the anisotropy ratio derived from $B_{c2\perp}$.
[c]obtained from $B_{c1\perp}$ and $B_{c2\perp}$.
[d]obtained from specific-heat data.
[e]from [212].
[f]from [227].
[g]from [228].

For β-(ET)$_2$I$_3$ similar values of $\xi_\parallel = 63$ nm and $\xi_\perp = 2.0$ nm are obtained with the critical field data of [191]. Here as well as above, an isotropic $B_{c2\parallel}(0) = 2.5$ T within the ET planes was assumed. For this superconductor, therefore, the simple paramagnetic limit of ~ 2 T is also exceeded.

The parameters for κ-(ET)$_2$Cu(NCS)$_2$ in Table 2.2 are obtained from recent magnetization data in higher fields for B_{c2} [189] and lower fields for B_{c1} [203]. In an earlier experiment a lower critical field $B_{c1\perp} = 16$ mT was extracted in a somewhat different method from the magnetization curves [204]. In a numerical approximation demagnetization effects have been tried taking into account B_{c1} as the fitting parameter [205]. This resulted in a fairly straight line of B_{c1} vs T. However, here also B_{c1} seems to be below the sensitivity limit from $T_c \approx 9\,\mathrm{K}$ down to $\sim 7.5\,\mathrm{K}$.

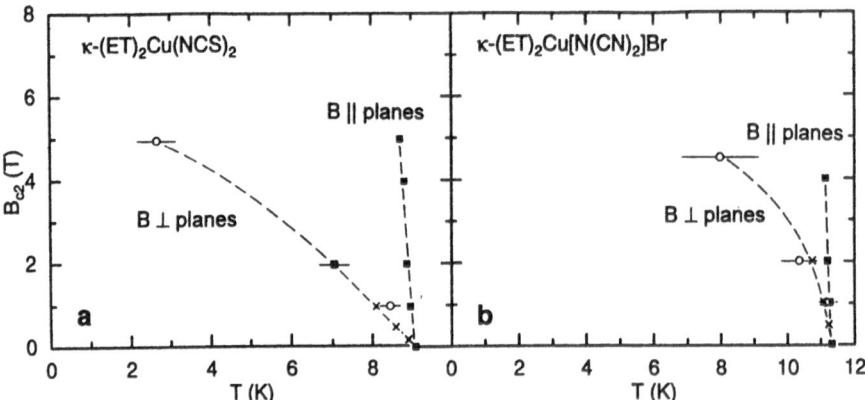

Fig. 2.30. Temperature dependence of the upper critical fields of (a) κ-(ET)$_2$-Cu(NCS)$_2$ and (b) κ-(ET)$_2$Cu[N(CN)$_2$]Br extracted from magnetization measurements. From [189]

The high-field magnetization data of [189] have been scaled using a 3D scaling form [188]. The analogous 2D scaling, however, works equally as well. From that $T_{c2}(B)$, which is the critical temperature at B_{c2}, can be deduced. The resultant upper critical fields for κ-(ET)$_2$Cu(NCS)$_2$ and κ-(ET)$_2$Cu[N(CN)$_2$]Br are shown in Fig. 2.30 [189]. The values for $B_{c2\parallel}$ are lower bounds because small misalignments of the sample could not be excluded. This, however, would result in drastic reductions of B_{c2} as is obvious from Fig. 2.27.

From the initial slopes of dB_{c2}/dT at T_c the GL coherence length was estimated from

$$dB_{c2\perp}/dT = \frac{\phi_0}{2\pi\xi_\parallel^2 T_c}, \qquad (2.20)$$

and $\xi_\perp/\xi_\parallel = B_{c2\perp}/B_{c2\parallel}$. Thereby it is assumed that $\xi \propto (T - T_c)^{-1/2}$ is valid for the whole temperature range. How far this is justified is questionable. In Fig. 2.30 it is obvious that at least for $B_{c2\perp}$ the linear temperature relation does not hold to the lowest temperature. Especially for κ-(ET)$_2$Cu[N(CN)$_2$]Br a strong deviation from expected linearity near T_c is found. In addition, with the initial slopes of $dB_{c2\parallel}/dT = -13\,\mathrm{T/K}$ for $X = $ Cu(NCS)$_2$ and $-21\,\mathrm{T/K}$ for $X = $ Cu[N(CN)$_2$]Br the critical fields at zero would be 120 T and 240 T, respectively. These values are far larger than the Pauli limits. However, as long as there are no high-field data, the critical fields close to $T = 0$ and the given values of ξ have to be regarded with caution.

The critical-field values of κ-(ET)$_2$Cu(NCS)$_2$ obtained in [189] exceeded those of [206] by a factor of ~ 1.7 where a more elaborate formula to fit the

M vs T data was used. In any case, magnetization data differ significantly from previous B_{c2} data extracted from resistivity transitions without taking fluctuations into account [207]. The values for κ-(ET)$_2$Cu[N(CN)$_2$]Br agree within 20% with other literature data [199]. In Table 2.2 $B_{c2\perp}$ was estimated by smoothly extrapolating $B_{c2}(T)$ to $T = 0$. $B_{c2\parallel}$ was then obtained with the known anisotropy of the initial slope. This yields, of course, somewhat smaller critical fields and, therefore, larger ξs than in [189, 199, 206] where $B_{c2\perp}(T)$ was linearly extrapolated to $T = 0$.

A thermodynamic quantity not very often measured for organic superconductors is the specific heat, C. Usually the crystal sizes are rather small and consequently a high sensitivity of the apparatus is needed. In most experiments, therefore, an assembly of many pieces of material is necessary to gain better resolution. In addition, the jump of C at T_c is expected to be rather small especially for compounds with higher transition temperatures because of the comparatively large lattice contribution to C owing to the low electron density and the low vibrational frequencies.

Nevertheless, it was quite surprising that in the first specific-heat experiment, the measurement of β-(ET)$_2$I$_3$ down to 0.7 K, no signs of an anomaly at T_c could be found [208]. This apparent contradiction to the Meissner measurements [181] might be due to a spread of T_c in the samples and a lack of resolution in the measurement. Later C experiments of other ET superconductors could well resolve the jump ΔC at T_c and a temperature dependence of C in reasonable agreement with BCS theory. From some of these experiments, however, a tendency towards strong coupling was concluded [209, 210, 211, 212].

A recent specific-heat measurement is shown in Fig. 2.31 for κ-(ET)$_2$I$_3$ [198]. At $T_c = 3.4$ K (3.5 K with magnetization measurements at the same crystal) a clear anomaly in C/T^3 vs T can be seen. The height of the jump at T_c is $\Delta C \approx 103$ mJ/molK. In a small field applied perpendicular to the ET planes the anomaly of C becomes smaller and much broader. In an overcritical magnetic field of $B_\perp = 0.5$ T (not shown here) the normal-state specific heat was measured. Besides the usual linear electronic and cubic Debye specific heat a hyperfine contribution at low temperatures and an appreciable T^5 phononic term had to be taken into account. Therefore, below 5 K C was fitted by

$$C = a_{hf}T^{-2} + \gamma T + \beta T^3 + \delta T^5, \tag{2.21}$$

with $a_{hf} = (0.77 \pm 0.05)$ mJK/mol, $\gamma = (18.9 \pm 1.5)$ mJ/molK^2, $\beta = (10.3 \pm 1)$ mJ/molK^4, and $\delta = (1.03 \pm 0.2)$ mJ/molK^6. The latter contribution results from low-energy vibrational excitations. The temperature dependence of these excitations up to higher temperatures can be approximated by $N_E = 1.4 \times 10^{24}$ Einstein modes per mol with a characteristic temperature of $\Theta_E = 28$ K estimated from

$$C_E = N_E k_B \left[\frac{\Theta_E}{T}\right]^2 \frac{\exp(\Theta_E/T)}{[\exp(\Theta_E/T) - 1]^2}. \tag{2.22}$$

With Raman scattering experiments close to the corresponding Einstein-mode energy ($\sim 2.3\,\text{meV}$ or $\sim 19.4\,\text{cm}^{-1}$) optical modes probably due to librations of the ET molecules were observed [213].

From the linear coefficient, γ, a value of $\Delta C/\gamma T_c = 1.6 \pm 0.2$ is obtained which is only slightly larger than the BCS prediction of 1.43. However, when plotting the difference between the specific heat in the superconducting and normal state a clear deviation from the BCS behavior (solid line in Fig. 2.31b) [214] is seen. The measured C below T_c falls off more steeply than predicted. This is an indication for strong coupling.

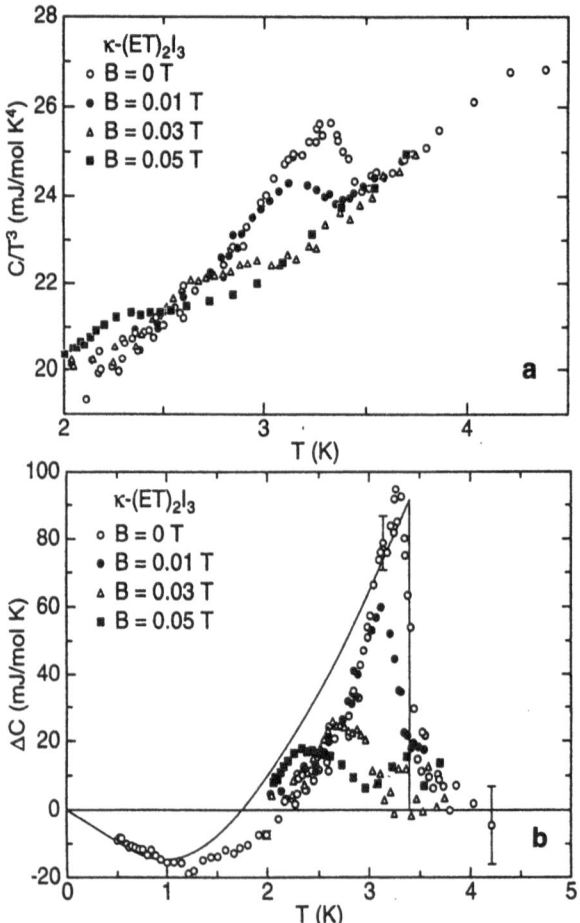

Fig. 2.31. (a) Specific heat of κ-(ET)$_2$I$_3$ divided by T^3 against temperature. (b) Specific-heat difference between the superconducting and the normal state. The solid line is a BCS fit for $T_c = 3.4\,\text{K}$

In an earlier specific-heat investigation of κ-(ET)$_2$Cu(NCS)$_2$ the data appeared to follow the BCS behavior perfectly [212]. A closer look at the BCS fit curve used in [212] revealed, however, a discrepancy to the correct BCS dependence. Obviously the authors have calculated their fit curve incorrectly (see also the discussion in [198]). The actual temperature dependence of ΔC for both κ-(ET)$_2$I$_3$ and κ-(ET)$_2$Cu(NCS)$_2$ is very similar which shows the common nature of the superconductivity in these organic materials.

For most other organic superconductors where a jump in C at T_c could be observed, the height of the jump was compatible with strong coupling only, too. For κ-(ET)$_2$Cu[N(CN)$_2$]Br, however, there exists a slight discrepancy between two experiments. One group [209] found a broad anomaly at T_c, while the other group which claimed a higher experimental resolution could not resolve any sign of a discontinuity [215]. The expected relative effect of the C jump at $T_c = 11.4$ K is of the order of 1–2% due to the large vibrational background. The Debye temperature of $\Theta_D = 218$ K for κ-(ET)$_2$I$_3$ is comparable to literature data for other organic metals [210, 211, 212]. The γ value is slightly lower than obtained from experiments for other materials [211], but in excellent agreement with the 2D free-electron model with an effective mass of $3.9\,m_e$ as obtained by dHvA experiments (see Sect. 4.2.3).

From the specific-heat data the thermodynamical critical field B_{cth} can be estimated [102, 194]. For κ-(ET)$_2$I$_3$ this results in $B_{cth} \approx 17$ mT. B_c can also be calculated from the upper and lower critical fields via

$$B_{cth} = \frac{B_{c2}}{\kappa\sqrt{2}} = \frac{B_{c1}\sqrt{2}\kappa}{\ln\kappa}, \tag{2.23}$$

where κ itself can be obtained by the implicit relation (see also (2.9))

$$\frac{B_{c1}}{B_{c2}} = \frac{\ln\kappa}{2\kappa^2}. \tag{2.24}$$

The results are shown in Table 2.2. The value $B_c = 20$ mT obtained in this way is in reasonable agreement with the critical field derived from the C measurements regarding the error bars involved in the derivation of both values. There is, however, a large discrepancy in the value of B_{cth} obtained from the critical fields for parallel orientation. The reason for this unusual behavior is unclear. It might be an indication for the decoupling of the 2D superconducting layers. Another possible explanation would be an upper critical field $B_{c2\parallel}$ which is reduced due to the paramagnetic limit [183].

The experimental results discussed so far can be explained reasonably well by the conventional theory for anisotropic strong-coupling type-II superconductors. In contrast, a controversy about the interpretation of NMR results exists. In these experiments a characteristic feature for different ET superconductors is the enormous enhancement of the spin-lattice relaxation rate, $1/T_1$, at approximately $0.5\,T_c$ [216, 217, 218]. As an example, the result of T_1^{-1} at $B = 0.61$ T for the "high-T_c" phase of β_H-(ET)$_2$I$_3$ ($P = 1.6$ kbar) is shown in Fig. 2.32. NMR experiments were highly important in establishing the BCS theory through the observation of the coherence peak at $\sim 0.85\,T_c$

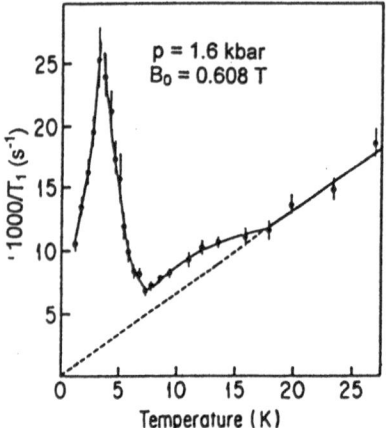

Fig. 2.32. Temperature dependence of the proton NMR spin-lattice relaxation rate of β_H-$(ET)_2I_3$. From [216]

with an increase in T_1^{-1} of approximately a factor two in clean materials [118]. This peak should be observed when the superconducting gap opens and s-wave pairing is dominant. The extreme T_1^{-1} enhancement by roughly a factor of 10 seen in Fig. 2.32 and similar results found in many other organic superconductors was, therefore, interpreted as being an indication for unconventional superconductivity in these compounds. Many theories were proposed to explain the unusual effect by critical fluctuations near the phase transitions [216], by a SDW or structural transition, or by flux-line lattice melting [217].

In a recent experiment the spin-lattice relaxation rate of one 2.2 mg κ-$(ET)_2Cu[N(CN)_2]Br$ single crystal was measured as a function of sample orientation with respect to the field [218]. The result is shown in Fig. 2.33. The main result of this investigation is the strong reduction of T_1^{-1} in a narrow angular region ($\sim 4°$) for the field applied close to the direction parallel to the ET planes. This peculiar behavior cannot be described by a conventional electronic relaxation mechanism. In addition, because the dip of T_1^{-1} is seen only in the superconducting state a connection to the flux-line vortices in the Shubnikov state is highly suggestive.

The fluxoids in 2D superconductors are relatively free to move as long as the field is not parallel to the ET layers where the flux lattice is pinned by the crystal structure itself, so-called intrinsic pinning. This effect has been seen by resistivity measurements in $YBa_2Cu_3O_{7-\delta}$ [219] and should be visible in κ-$(ET)_2Cu[N(CN)_2]Br$, too, where $\xi_\perp$ is also less than the interlayer distance (see Table 2.2). Therefore, it is proposed that the large proton relaxation rates observed below T_c in organic superconductors are mainly due to fluxoid dynamics [218]. In the organic metals studied the ^1H nucleus has a relatively weak coupling to the conduction electrons and, consequently, the large relaxation rates due to the moving flux lines are dominant. In contrast, in the

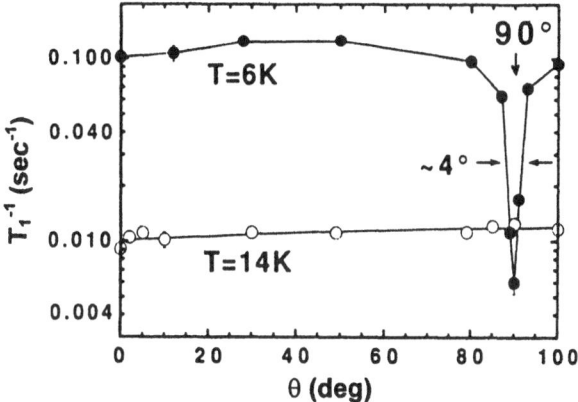

Fig. 2.33. Angular dependence of the ^{1}H NMR spin-lattice relaxation rate for κ-(ET)$_2$Cu[N(CN)$_2$]Br. From [218]

cuprate superconductors, which have already very high relaxation rates due to the conduction electrons, this behavior is masked.

A further puzzling and controversial issue is the temperature dependence of the magnetic penetration depth, $\lambda(T)$. This quantity measures the length over which magnetic fields penetrate into a superconductor. The temperature dependence of λ gives some information of the order-parameter symmetry in superconductors. For conventional s-wave BCS-like superconductors at $T < T_c$, $\lambda(T)$ should decrease exponentially towards lower temperatures. If, however, the energy gap topology is anisotropic a power-law dependence on T is expected.

The experimental situation is inconclusive and sometimes even the same experimental techniques used by different groups give contrary results. Especially for the compounds κ-(ET)$_2$Cu(NCS)$_2$ and κ-(ET)$_2$Cu[N(CN)$_2$]Br many different techniques have been employed to measure $\lambda(T)$. Evidence for non BCS-like behavior has been obtained by complex ac susceptibility [220], radio-frequency penetration depth [221], muon spin relaxation (μSR) [222], and microwave surface impedance measurements [223]. In contrast, results consistent with conventional BCS theory, sometimes revealing a tendency towards strong coupling, are reported for measurements of the μSR [224], microwave surface impedance [225, 226], and dc magnetization [227].

Figure 2.34 illustrates these serious experimental problems. In the upper part (a) the magnetic penetration depth of κ-(ET)$_2$Cu(NCS)$_2$ measured by ac susceptibility [220] and in the lower part (b) the same quantity obtained by dc magnetization [227] is shown. The clear discrepancy of the behavior of $\lambda(T)$ at low temperatures is obvious. In the ac-susceptibility experiment consistent with [221, 222, 223] a T^2 variation is found. This non-exponential non-BCS dependence can be explained with energy-gap nodes of several different topologies [222]. On the other hand, the data of [227] could very well be described by conventional weak-coupling theory in the clean limit as shown

by the solid line in Fig. 2.34b. The broken lines represent model curves for
two triplet states denoted t_1 and t_4 after Ref. [222]. The same experiment for
κ-(ET)$_2$Cu[N(CN)$_2$]Br revealed BCS-like behavior as well, however, with a
tendency to strong coupling [228].

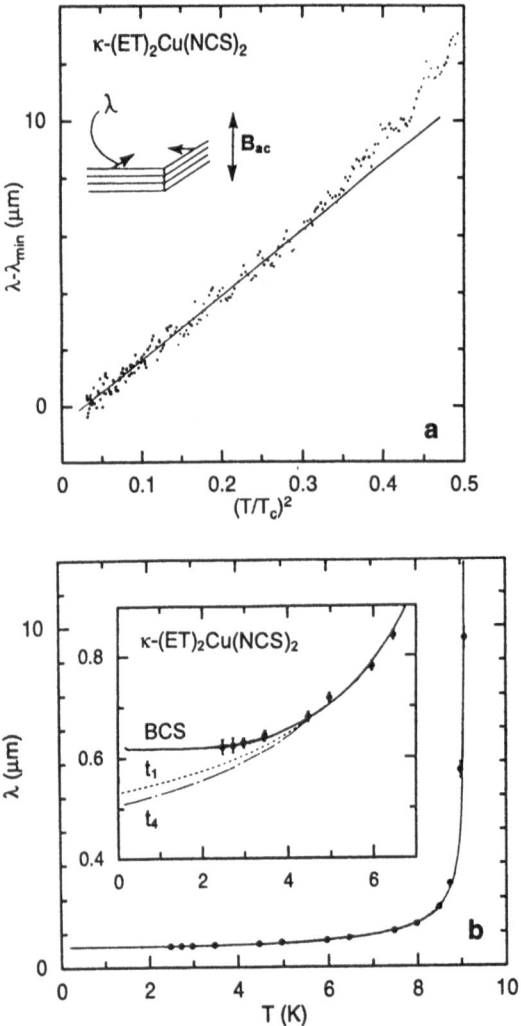

Fig. 2.34. Temperature dependence of the in-plane magnetic penetration depth
extracted from (a) ac-susceptibility and (b) dc-magnetization measurements. The
inset in (b) shows the low-T data in enlarged scale together with model curves for
BCS (solid line) and for two triplet states, t_1 and t_4 (broken curves). From [220]
and [227]

The inconclusive experimental picture concerning $\lambda(T)$ prevails in microwave experiments of [223] and [226], respectively, where the measured data

of both groups are essentially the same. The conclusions, however, depend on whether the analysis of the data starts from the usual BCS-like ansatz or not.

As a further test as to whether the behavior is conventionally supercon-ducting or not, the isotope effect of organic superconductors has been studied at length. From the simple BCS formula (2.6) it follows that T_c should in-crease with increasing Debye temperature which is to a first approximation proportional to $m^{-1/2}$ (see also Chap. 1). However, the results of isotope substitutions are rather inconclusive. Everything between the decrease of T_c with increasing isotope masses as for β_H-(ET)$_2$I$_3$ [210] and the opposite for κ-(ET)$_2$Cu(NCS)$_2$ [229] has been reported. The crucial point is the apprecia-ble variation of T_c between different batches and even of crystals within one batch. As a consequence in a recent experiment a statistical approach was chosen [230]. In addition, the identical solvents for the isotope-substituted salts have been used in this experiment of isotope substitutions in β-(ET)$_2$I$_3$, κ-(ET)$_2$Cu(NCS)$_2$, and κ-(ET)$_2$Cu[N(CN)$_2$]Br. Within the statistical accu-racy no conclusive shift of T_c has been observed. In addition, the expected isotope effect should be rather small (a few percent at most) because of the large number of atoms per unit cell and, therefore, large molar masses.

The absence of the isotope effect, on the other hand, is no direct evidence for an unconventional non-phonon mediated pairing mechanism. This neg-ative result is well known from a number of transition elements (see [102]) which is explicable by the more elaborate McMillan formula [231]

$$T_c \propto \omega_D \exp\left(-\frac{\lambda+1}{\lambda - \mu^*(1 + \lambda\langle\omega\rangle/\omega_D)}\right), \qquad (2.25)$$

where $\omega_D = k_B\Theta_D/\hbar$ is the Debye frequency, μ^* describes the Coulomb in-teraction, and $\langle\omega\rangle$ is an average over the phonon frequencies. It is clear from this equation that the relation $T_c \propto m^{-1/2}$ is certainly not straightforward since ω_D is standing both in front as is also in the exponential function with different effect on T_c.

Deviations from the expected isotope effect might be further expected from Coulomb correlations in the electronic system. These correlation effects are expected to be present in low-dimensional systems with low charge-carrier densities. Some experimental indications for a strong electron–electron inter-action in the ET materials has been found which will be discussed in Sect. 4.2.

Similar to the 1D organic metals, the superconductivity in ET salts is highly sensitive to disorder. This is obvious from the low-pressure phase of β-(ET)$_2$I$_3$ where T_c is suppressed by the incommensurate superstructure al-ready mentioned, or from the absence of superconductivity in β-(ET)$_2$I$_2$Br due to anion disorder [128, 131]. Systematic investigations of the composition dependence of T_c in β-(ET)$_2$(I$_3$)$_{1-x}$(IBr$_2$)$_x$ showed a rapid suppression of su-perconductivity for concentrations only slightly away from $x = 1$ and $x = 0$ [232]. This is contrary to the behavior found usually in superconductors and was interpreted as a further indication of an enhanced Coulomb repulsion due to the randomness [233].

Measurements of the superconducting energy gap by point-contact spectroscopy [234], vacuum tunneling [235], and scanning tunneling microscopy [236] all showed structures at significantly higher energies than predicted by weak-coupling BCS theory. The interpretation of these data is still rather preliminary. Claims that the structure is due to a large gap resulting from extremely strong coupling seem to be rather inconsistent with respect to the experimental results discussed before. A final conclusive statement, therefore, is not possible and further experiments would be highly desirable.

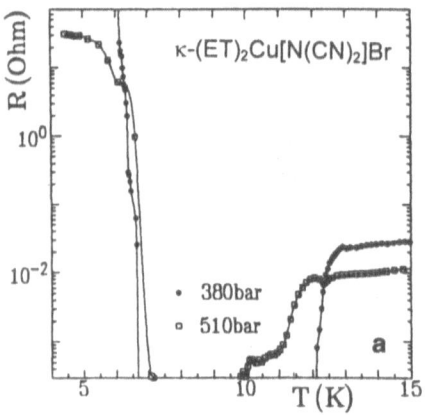

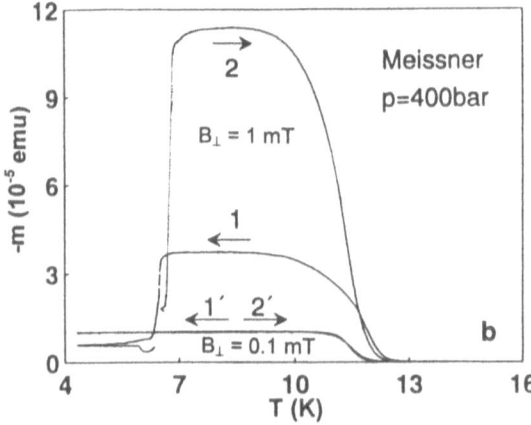

Fig. 2.35. Temperature dependence of (**a**) the resistivity and (**b**) dc magnetization of κ-(ET)$_2$Cu[N(CN)$_2$]Cl for different applied pressures showing the reentrant behavior. From [238] and [242]

A rather peculiar superconducting behavior was found in κ-$(ET)_2$Cu-$[N(CN)_2]$Cl, the organic 2D superconductor with the highest T_c to date. This material shows a remarkable phase diagram with a rich variety of phases. It becomes superconducting at ~ 12.8 K only under an applied pressure of above ~ 300 bar [16, 237]. Later experiments revealed that below this pressure the non-metallic and the superconducting states coexist [238]. This non-metallic state is very complex and below ~ 45 K seems to be antiferromagnetic. Upon further cooling below ~ 22 K weak ferromagnetism has been observed [239]. Proton NMR measurements have proven a commensurate antiferromagnetic ordering, however only below ~ 27 K [240]. The weak ferromagnetic moment is thought to be due to a canting of the ordered spins. Further systematic investigations under different pressures revealed an even more exotic feature, namely a reentrant behavior of the non-metallic state with decreasing temperature and applied moderate pressure [238, 241]. The resistivity and magnetization measurements [242] are shown in Fig. 2.35. The observed Meissner effect clearly confirms that the superconductivity in the intermediate temperature region is a bulk effect. In addition, it was found that the already complex phase diagram is highly sensitive to the cooling rate and the thermal and magnetic history of the sample [241, 243]. All these effects reflect the subtle balance between electronic interactions leading to ferromagnetism, antiferromagnetism, or superconductivity.

Reentrant superconductivity is known from so-called ferromagnetic superconductors like $ErRh_4B_4$ where the ferromagnetic transition at lower temperatures actually destroys superconductivity [244]. However, the exact magnetic structure in κ-$(ET)_2$Cu$[N(CN)_2]$Cl is still unknown and the origin of the peculiar behavior is at present subject of speculations only.

3. Magnetic Oscillations in Metals

Although the first observation of oscillatory effects in the magnetization and resistivity of bismuth dates back to 1930 [245, 246] the rigorous theoretical understanding of this FS investigation technique was only accomplished in the 1950s and 1960s. In the following with the improvement of experimental equipment, especially superconducting magnets with large and homogenous fields as well as the availability of purer samples, dHvA and SdH measurements contributed enormously to the understanding of electronic properties in metals. "Fermiology", that is the study of the electronic topology in metals both theoretically and experimentally, has become an established and important tool in solid state physics.

With the discovery of organic metals and superconductors one of the main questions was what the electronic structure looks like in the different compounds. In 1D materials the usual dHvA and SdH effect cannot be used as a tool for probing the FS due to the lack of closed electron orbits. Recently, however, it was shown that with a different approach, namely by measuring the angular dependence of the resistance in large magnetic fields and at low temperatures some topological aspects of the 1D FS can be extracted. The 2D organic metals, on the other hand, are ideally suited for dHvA and SdH studies. Since the first reports of SdH oscillations in ET compounds [29, 30] the number of investigations of magnetic quantum oscillations has increased steadily.

It is beyond the scope of this monograph to give an explicit derivation of the theory of magnetic oscillations in metals. The reader is referred to several reviews [247, 248] and, above all, to Shoenberg's book [249]. Here only the principal derivation of the theoretical description for the oscillations of the magnetization (dHvA effect) and some general remarks concerning the SdH effect are presented. Section 3.3 deals with the actual theoretical understanding of the angular dependence of magnetoresistance oscillations (AMRO), an experimental technique that has become increasingly important for the extraction of 1D and 2D FS topologies.

3.1 de Haas–van Alphen Effect

3.1.1 Basic Theory

The basic semi-classical equation of a magnetic field B acting on an electron[1] in a metal is[2]

$$\hbar\dot{k} = -e v \times B, \tag{3.1}$$

where k is the wave vector of the electron, v is the electron velocity, and e is the (positive) electronic charge. The velocity is given by the energy dispersion, i. e.,

$$v = \frac{1}{\hbar}\mathrm{grad}_k\epsilon. \tag{3.2}$$

From these two equations it can easily be seen that the electrons are moving on an orbit in k space which is given by a constant energy surface perpendicular to B. The angular frequency with which the so-called cyclotron orbit is traced is given by the cyclotron frequency $\omega_c = eB/m_c$, where the cyclotron mass is defined by

$$m_c = \frac{\hbar^2}{2\pi}\left(\frac{\partial a}{\partial\epsilon}\right)_\kappa. \tag{3.3}$$

$(\partial a/\partial\epsilon)_\kappa$ is the derivative of the cyclotron orbit area with respect to energy keeping the component κ of k along B constant.

In order to understand the occurrence of dHvA oscillations one has to take into account the quantization of the electron motion. The Bohr–Sommerfeld quantization rule for an electron in a magnetic field is

$$\oint (\hbar k - e A)\mathrm{d}R = (n + \gamma)2\pi\hbar, \tag{3.4}$$

where A is the vector potential of B, n is an integer, and γ is a phase which usually is close to $1/2$ (exactly $1/2$ for a parabolic band). Thus the electron motion is quantized and only discrete orbits are allowed. The enclosed area, a, of the orbits is given by Onsager's quantum condition

$$a = \left(n + \frac{1}{2}\right)2\pi eB/\hbar. \tag{3.5}$$

The allowed states in k space are all lying on so-called Landau levels or Landau tubes. The annular cross-sectional area between neighboring Landau levels is $\Delta a = 2\pi eB/\hbar$. All the electron states in k space which without fields

---

[1] The term "electron" should be read here as a wave packet of Bloch states centered at k, r with an extension Δk and Δr such that $\Delta k \ll k_F$ and $\Delta k\Delta r > 1$ imposed by the uncertainty principle.

[2] Here and in the following it is assumed that the difference between B and $\mu_0 H$ is negligible. If, however, magnetic interaction effects come into play one has to distinguish between both. This is briefly discussed in Sect. 4.2.2 and in more detail in [249].

lie between two neighboring tubes become degenerate in field and condense onto the Landau levels. If now B increases the cross-sections of the Landau tubes equally increase and eventually one tube will cross the highest occupied energy state given by ϵ_F, or equivalently by k_F or a_F (= the extremal area of the Fermi surface perpendicular to B). At this field the electrons from the outermost occupied tube have to redistribute on lower Landau levels. This, in consequence, gives rise to a change in the free energy of the electronic system. Thus, all derived quantities, in particular the magnetization M, the derivative of the free energy with respect to field, will show some kind of structure. From (3.5) it can be seen that this redistribution for fixed a_F is periodic in $1/B$.

The explicit form up to second order of the oscillatory part of $M(B)$ for a parabolic band was first given by Landau [250]. Taking into account phase smearing effects due to finite temperature, electron scattering, and electron spin finally resulted in the so-called Lifshitz–Kosevich formula [251] which is given here for one extremal area of the FS:

$$M \propto \left(\frac{B}{|\partial^2 a_F/\partial \kappa^2|}\right)^{\frac{1}{2}} \sum_{r=1}^{\infty} \frac{(-1)^r}{r^{3/2}} R_D R_T R_S \sin\left[2\pi r\left(\frac{F}{B}-\gamma\right)\pm\frac{\pi}{4}\right]. \quad (3.6)$$

This formula is valid only for a three-dimensional FS and describes the field-dependent oscillatory part of the magnetization caused by conduction electrons in a metal. The curvature factor $|\partial^2 a_F/\partial \kappa^2|^{-1/2}$ describes the curvature of the FS, i.e., the change of a_F parallel to B. The dHvA frequency F is directly proportional to a_F and is easily derived from (3.5), $F = \hbar a_F/2\pi e$. The unit of F is T. The damping factors R_D and R_T describe the phase smearing, i.e., the superposition of oscillations with different phases due to scattering of the electrons and oscillations of slightly varying F due to the finite temperature T, respectively. R_S takes into account the Zeeman splitting of the Landau levels in the magnetic field. For metals with FSs which have more than one extremal area a_F the total oscillatory M is simply the sum over all contributions each of which has the form (3.6) but with different parameters F, a_F, and R_i.

In the following the damping factors R_i will be described in more detail. At $T > 0$ the border between occupied and unoccupied states is no longer sharp but somewhat smeared according to the Fermi distribution function $f(\epsilon) = (1 + \exp[(\epsilon - \epsilon_F)/k_B T])^{-1}$. Therefore, the electrons on the outermost Landau tube do not all redistribute at exactly the field values corresponding to a_F but rather over a temperature dependent field range. The resulting damping factor is

$$R_T = \frac{\alpha r \mu_c T/B}{\sinh(\alpha r \mu_c T/B)}, \quad (3.7)$$

where $\alpha = 2\pi^2 k_B m_e/e\hbar = 14.69\,\text{T/K}$ and μ_c is the effective cyclotron mass in relative units of the free electron mass m_e. For large arguments $(\alpha r \mu_c T/B \geq 2)$ the above expression can be approximated by

$$R_T \approx \frac{2\alpha r \mu_c T}{B} \exp(-\alpha r \mu_c T/B). \tag{3.8}$$

To illustrate the effect of finite temperatures on the dHvA amplitude further, Fig. 3.1 shows an actual measurement of the dHvA effect in κ-$(ET)_2I_3$ for different temperatures. This organic superconductor has a simple FS so that for the chosen field and temperature range only one extremal orbit is dominant (see Sect. 4.2.3). With increasing temperature the strong decrease of the oscillating magnetization is clearly seen. From this dependence, the cyclotron effective mass, μ_c, can be extracted either by fitting the relation

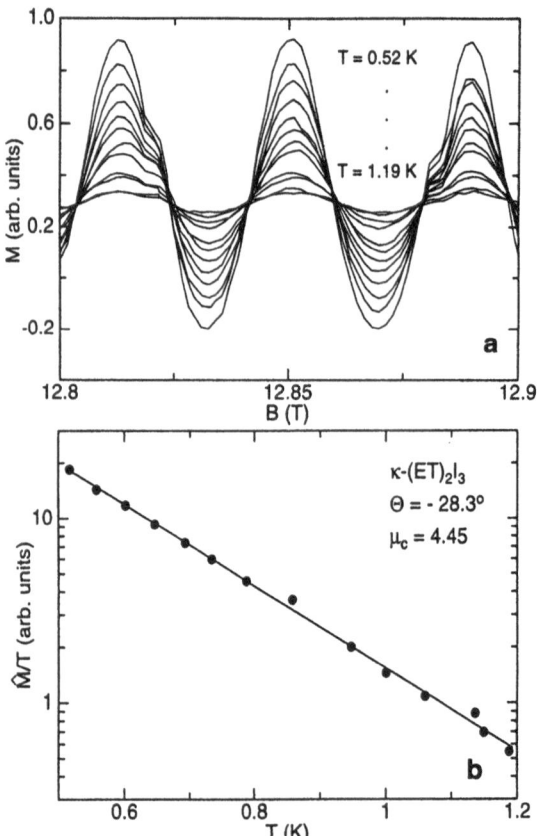

Fig. 3.1. (a) dHvA oscillations of κ-$(ET)_2I_3$ for different temperatures between 0.52 and 1.2 K in a narrow field range. (b) Temperature dependence of the dHvA amplitude divided by T with the fit according to (3.8)

$\hat{M}/T \propto 1/\sinh(\alpha\mu_c T/B)$ to the measured amplitude of the oscillating signal, $\hat{M}$, at fixed B or by linear regression of $\ln(\hat{M}/T)$ vs T using (3.8). A plot of the latter is shown in Fig. 3.1b. The slope gives $\alpha\mu_c/B$ and, therefore, directly μ_c. For the example shown with $\mu_c = 4.45$ the deviations from the exact formula (3.7) are less than 0.7% over the whole temperature range and can be neglected.

The Dingle reduction factor, R_D, describes the broadening of the otherwise sharp Landau levels due to scattering of the conduction electrons. The usual parameter which describes this scattering is the relaxation time τ averaged over one cyclotron orbit [252]. This effect leads to a reduction factor similar to (3.8) for finite temperatures. As a useful parameter the so-called Dingle temperature

$$T_D = \hbar/2\pi k_B \tau \tag{3.9}$$

is introduced. Thus, the Dingle factor can be written as

$$R_D = \exp(-\alpha \tau \mu_c T_D/B). \tag{3.10}$$

Here it is assumed that τ is independent of energy [249]. This means that the elastic and inelastic scattering process within a sample has almost the same effect as an increase of the measuring temperature. An additional damping may result from sample and field inhomogeneities. This yields a damping factor essentially equal to (3.10) and will not be discussed further here.

With the knowledge of μ_c from the temperature dependence of the dHvA amplitude it is possible to extract the Dingle temperature, T_D, of the sample investigated. Thereby, the field dependence of the dHvA amplitude of the particular extremal area has to be analyzed. From (3.6), (3.7), and (3.10) the relation $\hat{M}\sqrt{B} \propto \exp(-\alpha \tau \mu_c T_D/B)$ is obtained. Therefore, by a plot of $\ln(\hat{M}\sqrt{B})$ vs $1/B$ the Dingle temperature can be extracted by linear regression. The value of T_D is, of course, strongly sample dependent and may vary from crystal to crystal. For organic metals values usually around 0.5–3 K have been found, revealing good crystal quality. Hopefully, further improvements in the electrocrystallization technique will result in even better crystals.

Finally, the factor R_S takes into account that the degeneracy of the energy levels is lifted by the Zeeman splitting and that two sets of Landau levels evolve. The energy difference between the split levels is

$$\Delta \epsilon = g\mu_B B, \tag{3.11}$$

where $\mu_B = e\hbar/2m_e$ is the Bohr magneton and g is the spin g-factor which for free electrons is $g = g_0 = 2.0023 \approx 2$. The two sets of Landau tubes are passing the FS at different fields with a phase difference ϕ given by the ratio of the Zeeman splitting (3.11) averaged over a cyclotron orbit and the cyclotron energy $\hbar\omega_c$, namely

$$\phi = 2\pi \Delta\epsilon/\hbar\omega_c = \pi g m/m_e. \tag{3.12}$$

The superposition of the spin-up and spin-down oscillations shifted by ϕ results in the original dHvA amplitude multiplied by the spin-splitting reduction factor

$$R_S = \cos(\frac{1}{2}r\pi g\mu_b), \tag{3.13}$$

where $\mu_b = m/m_e$ is the band-structure effective mass in relative units. In contrast to μ_c this mass is not renormalized by the electron–phonon interaction, that is $\mu_c = \mu_b(1+\lambda)$ [249, 253, 254]. Therefore, under special conditions and when the g value is known λ maybe extracted from dHvA measurements (see Chap. 4).

3.1.2 Experimental Realization

Although in principle every experimental setup that measures the field dependence of the magnetization or a derivative of M could be used to observe the dHvA effect, usually an apparatus with a very high sensitivity has to be designed to resolve the oscillations. An overview of the different experimental techniques is given in Ref. [249]. Two main realizations used to detect dHvA oscillations in organic superconductors are the torque and the modulation-field method.

(i) The torque, T, acting on a sample with anisotropic FS is given by

$$T \propto -\frac{1}{F}\frac{\mathrm{d}F}{\mathrm{d}\Theta}MB, \qquad (3.14)$$

where M is given by (3.6). Therefore, the observation of a torque signal requires that the extremal area of the FS should change with angle. As will be shown later explicitly, the angular dependence of F for the 2D organic metals follows mostly the relation $F \propto 1/\cos\Theta$ which means that $\mathrm{d}F/\mathrm{d}\Theta$ is zero for $\Theta = 0$, i. e., for the field perpendicular to the conducting planes. Thus, measurements of the torque in 2D metals can only be done for tilted samples. The sensitivity can be very high with suitable feedback systems and sophisticated torque detection methods [255]. In addition, the absolute value of M can be measured simultaneously with the oscillatory part.

(ii) The highly sensitive field modulation technique is analogous to the usual mutual inductance method for ac-susceptibility measurements. In addition to the quasi-static field B a small periodic field $b(t) = b_0 \cos\omega t$ is superimposed which we assume to be parallel to B.[3] Therefore, the oscillatory magnetization is given by

$$M = \sum_{r=1}^{\infty} D_r(B,T,T_\mathrm{D}) \sin[2\pi r F/(B + b_0 \cos\omega t) + \gamma \pm \pi/4], \qquad (3.15)$$

where $D_r(B,T,T_\mathrm{D})$ comprises all the prefactors and damping terms discussed above. Now, with the assumption $b(t) \ll B$ and by considering only one term in the above sum (3.15) can be written to lowest order in $b(t)/B$

$$M = D_r \sin[2\pi r F/B - \Lambda \sin\omega t + \gamma \pm \pi/4], \qquad (3.16)$$

where $\Lambda = 2\pi r F b_0/B^2$.

---

[3] For a more general discussion see [248].

In a pick-up coil system the induced voltage $u_{\text{ind}} \propto \mathrm{d}M/\mathrm{d}t = \mathrm{d}M/\mathrm{d}B \cdot \mathrm{d}B/\mathrm{d}t$ is detected. With a little mathematics the pick-up signal can be calculated and is given by

$$u_{\text{ind}} = \Omega D_r \sum_{n=1}^{\infty} n\omega \cos[2\pi r F/B - n\pi/2 + \gamma \pm \pi/4] J_n(\Lambda) \sin n\omega t, \quad (3.17)$$

where $J_n(\Lambda)$ is the nth order Bessel function of the first kind with argument Λ, and Ω is a proportionality factor of the pick-up coil system.

The principal advantage of the modulation field method is the possibility of using a phase sensitive detector which can select higher harmonics ($n > 1$) of the pick-up signal. Hence, drift of the background magnetization can be easily suppressed. In addition, this technique gives the possibility of suppressing an especially strong dHvA frequency so that weaker oscillations may become visible. This can be achieved by adjusting the modulation field amplitude b_0 to a value so that Λ corresponds to one of the Bessel function zeros.

Usually for the observation of dHvA oscillations in organic metals temperatures below a few K down to the mK range are necessary. Therefore, the detection unit is inserted into a ^{3}He cryostat or a ^{3}He/^{4}He dilution refrigerator. Often the sample is directly immersed in the liquid and can be rotated *in situ* around one or even two axes. Thus, the FS can be mapped out more easily without remounting of the sample.

If a metal has more than one extremal area of the FS the usual way to deconvolute the different dHvA frequencies is done by a Fourier transformation with respect to $1/B$. This is, in addition, the easiest way to resolve higher harmonics r of the dHvA signal.

3.2 Shubnikov–de Haas Effect

Magnetic quantum oscillations in bismuth were first observed in the field dependence of the electrical resistivity by Shubnikov and de Haas [246] shortly before the dHvA effect was discovered. Usually, however, the SdH effect is weak and hard to observe except in semimetals, like bismuth, and semiconductors.

Although the theory of the SdH effect [256], which deals with the detailed problem of electron scattering in a magnetic field, is quite complicated a qualitative explanation is possible by virtue of a simple argument [257]. The probability for an electron to scatter is proportional to the number of states into which it can be scattered. As discussed above, for a metal in a magnetic field the density of states at the Fermi level $N(\epsilon_{\mathrm{F}})$ will oscillate with the field and, therefore, the scattering probability and the electronic relaxation time τ will oscillate, too. It can be shown that the oscillatory part of the density of states, $\tilde{N}(\epsilon_{\mathrm{F}})$, has essentially the same analytic form as (3.6) with the same reduction factors. The order of magnitude of the relative conductivity oscillations $\tilde{\sigma}/\sigma$ is related to $\tilde{N}(\epsilon_F)/N_0(\epsilon)$ where N_0 is the steady density of states. In low-dimensional organic metals the density of states is low and the highly symmetric form of the FS causes a large curvature factor with a large fraction of the density of states that oscillates. In addition, the number of different FS sheets is usually very small, whereas in polyvalent metals many sheets are present which are all contributing to the total conductivity. For typical conditions in usual metals, therefore, the resistivity oscillations are less than 10^{-4}. In contrast, the signal can become huge (several 100% of the resistivity at $B = 0$) in organic metals.

Indeed, most observations of magnetic quantum oscillations in organic conductors have been made by resistivity measurements. Information regarding FS topology that can be obtained from SdH experiments is in principle the same as in dHvA experiments. The oscillation frequency F, the effective cyclotron mass m_c, the bare electron mass m, and also the Dingle temperature T_D are extractable in the same way as described for the dHvA effect. However, in resistivity experiments electron interference effects can produce additional structures in the measurements and complicate the interpretation of the data. In addition, the sometimes extremely large amplitude of the SdH oscillations can mimic anharmonic effects in apparent contradiction to (3.6). This is due to the fact that experimentally only the resistivity, R, can be measured. Therefore, in the case that $\tilde{\sigma}/\sigma$ is large $\tilde{R}/R$ will show strong anharmonicities. This effect might be the reason for the unusual SdH signals shown in Figs. 4.9 and 4.16. The correct extraction of $\tilde{\sigma}/\sigma$ out of the measured $\tilde{R}/R$ by the necessary matrix inversion is possible only if the Hall effect has been measured simultaneously. Since this has not been done for the SdH experiments discussed here the stated values of, e. g., μ_c and T_D have to be taken with some caution. In this respect the independent observation of dHvA oscillations is highly desirable in order to obtain reliable results of electronic parameters for strongly 2D metals. In Chap. 4 results of both techniques, SdH and dHvA experiments, will be presented and discussed.

3.3 Angular-Dependent Magnetoresistance Oscillations

In the course of investigating the electronic properties of organic metals in magnetic fields not only the above introduced well-known magnetic quantum oscillations appeared but also a variety of new oscillatory phenomena which are connected with the special low-dimensional FS topology. The microscopic explanation of these effects was developed mostly only after experimental observation and is still partially controversial. The topological effects, however, are understood and are now widely used to map out special cross-sections of the FS. The angular dependent magnetoresistance oscillations (AMRO) described in this section are not related to the usual SdH effect for anisotropic Fermi surfaces but they can be explained quasi-classically.

In perfectly 1D or 2D materials at least one transfer integral, which we assume here to be the z direction, is zero, $t_z = 0$. Generally, the application of an electric field redistributes the electrons in a way that a finite average velocity of the electrons, that is a current or a finite conductivity, is established in the electric field direction. If, however, the electric field is applied in the z direction no redistribution of states is possible and the average velocity (for $t_z \equiv 0$ even the velocity of each electron) is zero. In real materials t_z is nonzero and, therefore, an average velocity exists resulting in a finite conductivity.

In a magnetic field we have seen in the quasiclassical approximation from (3.1) and (3.2) that the electrons move on trajectories with constant energy perpendicular to the applied field. Depending on the special topology at certain angles of magnetic field direction these trajectories may average the electron velocity for all states in the z direction to zero. Consequently, at these particular angles the resistivity will increase by an amount that depends on the actual scattering time τ. If τ is too small the electron will be scattered out of the trajectory and, therefore, a finite velocity in the z direction will remain. In this section three different angular dependent magnetoresistance oscillations which might be all understood by this averaging effect will be discussed.

The first observation of AMROs was presented together with one of the first reports of SdH oscillations in the 2D organic metal β-(ET)$_2$IBr$_2$ [30]. Somewhat later the same effect was also found in Θ-(ET)$_2$I$_3$ [258]. The theoretical explanation was given shortly afterwards and is related to the slightly corrugated form of the cylindrical FS shown schematically in Fig. 2.22 in Sect. 2.3.2 [170, 171, 259]. In this section it was shown that for certain angles of the applied field given by (2.12) the extremal area of the FS cross-section perpendicular to $\boldsymbol{B}$ is constant. This, however, implies that the average velocity of the electrons in the z direction, that is perpendicular to the ET planes, tends to zero and consequently the resistivity has a maximum. As can be seen from (2.12), a plot of $\tan \Theta_n$ vs the nth maximum gives $\pi/c'k_F$. This result was obtained for a circular basal plane of the FS cylinder. If, however, the basal plane has arbitrary form and if the direction vector $\boldsymbol{h}$ of the interlayer transfer integral has an in-plane component $\boldsymbol{u}$, (2.12) is modified and the result is given by [172][4]

$$\tan \Theta_n(\varphi) = \frac{\left[\pi \left(n \mp \frac{1}{4}\right) + \left(k_\parallel^{\max}(\varphi) \cdot \boldsymbol{u}\right)\right]}{k_B^{\max}(\varphi) \cdot c'}, \tag{3.18}$$

where the signs $-$ and $+$ correspond to positive and negative Θ_n, respectively, $k_\parallel^{\max}(\varphi)$ is a Fermi wave-vector component in the ET plane whose projection

----

[4] In this reference the names of the angles φ and Θ are interchanged from the usage adopted here. To avoid confusion I stick to the same nomenclature throughout the book.

$k_B^{max}(\varphi)$ to the field rotation plane is maximal, and φ is an angle between the field rotation plane and a special direction (mostly the x direction) in the ET plane.

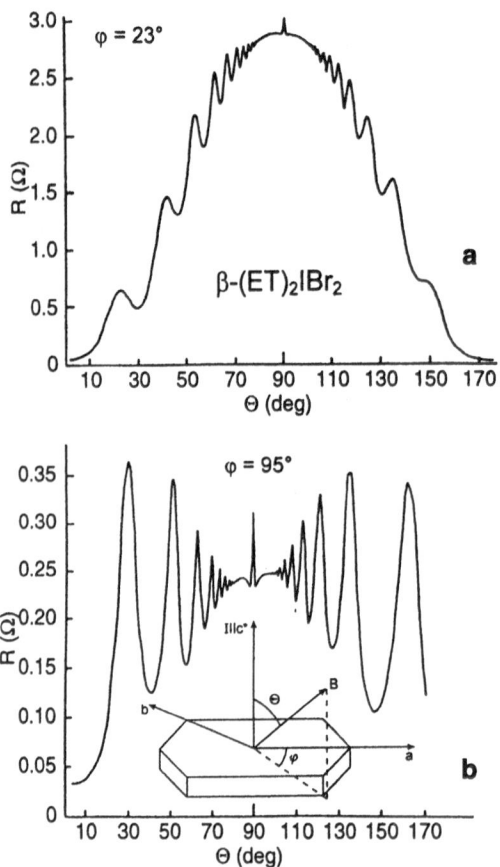

Fig. 3.2. Angular magnetoresistance dependence of β-(ET)$_2$IBr$_2$ for two different field rotation planes at $B = 14$ T. The angles Θ and φ are shown in the inset. From [172]

The actual experimental application of this technique is shown in Fig. 3.2 [172]. In this experiment the angular dependence of the magnetoresistance in a magnetic field of 14 T was measured for many different field rotation planes. From the slope of plots of $\tan \Theta_n$ vs n both u and $k_B^{max}(\varphi)$ could be obtained. The latter quantity can easily be related to $k_F(\varphi)$. The experimentally obtained points of the in-plane Fermi wave vector are shown in Fig. 3.3. Comparison of this result with the calculated FS in Fig. 2.17 indicates the excellent agreement between experiment and theory. The vector u was found to point perpendicular to the b axis. The angle between h and the z direction given by $\arctan(|u|/c')$ is $\sim 8°$ and corresponds approximately to the triclinic angle of c with respect to the z direction. Of course, this technique is so successful for the organic metals because of their simple band structure with only one or two bands at the Fermi level. The situation becomes much more complicated if more bands are involved in the charge transport.

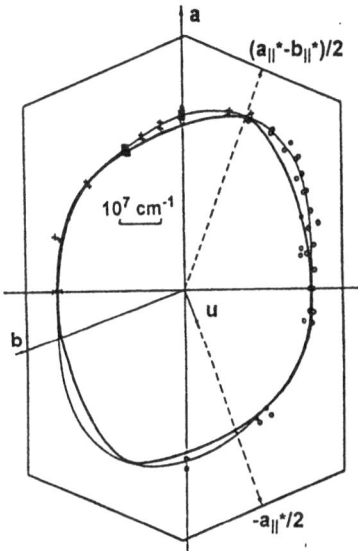

Fig. 3.3. From the angular dependence of the magnetoresistance obtained in-plane FS of β-(ET)$_2$IBr$_2$ (thick line). The thin line interpolates experimental points and represents $k_B^{\max}(\varphi)$. The magnitude of u is arbitrary. From [172]

A different situation occurs for the quasi-1D materials with a FS as shown in Fig. 2.3. Here the experimentally obtained information depends on which rotation plane of the field is chosen.

If the field is rotated in the bc plane, that is, perpendicular to the highly conducting a direction, at special angles weak structures in the resistance appear [260]. Figure 3.4 shows the experimentally observed angular dependence of the resistivity in the a direction of (TMTSF)$_2$ClO$_4$ for a field of 6 T rotated in the bc plane [260]. Some weak local minima in R marked by arrows are visible. In the lower part of this figure the second derivative of the resistance with respect to the angle is plotted. Now the distinct structures of $R(\Theta)$ at certain angles are clearly visible. These so-called "magic angles" or Lebed oscillations are given by the condition

$$\tan\Theta = \frac{p}{q} \cdot \frac{b\sin\gamma}{c\sin\beta\sin\alpha} - \cot\alpha, \qquad (3.19)$$

where p and q are integers and b, c, α, β, and γ are the lattice parameters of the crystal structure. For crystals with a rectangular lattice of higher symmetry this condition can be simplified to $\tan \Theta = (p/q) \cdot (b/c)$. The angles which satisfy this condition are labeled as p/q in Fig. 3.4. Here the formation of a superlattice in $(\mathrm{TMTSF})_2\mathrm{ClO}_4$ due to anion ordering with a doubling of the unit cell in b direction [70] has been taken into account.

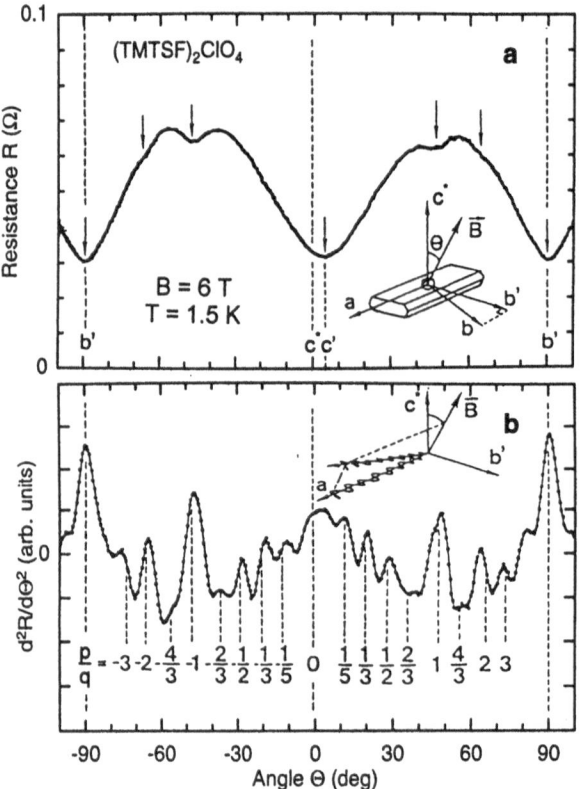

Fig. 3.4. (a) Angular dependence of the magnetoresistance of $(\mathrm{TMTSF})_2\mathrm{ClO}_4$ in 6 T and at 1.5 K. The field is rotated in a plane perpendicular to the highly conducting a direction. (b) Second derivative of the upper trace with the minima labeled as described in the text. From [260]

An explanation for this magic-angle effect has been given by Osada [261]. He considered the following x-direction linearized energy dispersion near the Fermi level

$$\epsilon(\boldsymbol{k}) = \hbar v_{\mathrm{F}}(|k_x| - k_{\mathrm{F}}) - \sum_{pq} t_{pq} \cos(pbk_y + qck_z), \qquad (3.20)$$

where t_{pq} represents the interchain transfer integral for the lattice vector $(0, pb, qc)$. Only t_{pq} with small p and q have finite values. If now a magnetic field is applied in a general direction in the bc plane the electron trajectory in the reduced first Brillouin zone eventually fills the whole corrugated FS area as shown schematically in Fig. 3.5a. The velocity components v_y and v_z take all possible values and thus average to zero. For special field orientations, however, the electron moves along the reciprocal lattice vector $K = pk_y + qk_z$. Hence the trajectory in the first Brillouin zone consists only of finite sets of lines (see Fig. 3.5b) so that the velocity perpendicular to a takes a limited number of values. Therefore, with the energy dispersion (3.20) the average velocity becomes non zero in this case and consequently dips of the transverse resistance components R_b and R_c are expected.

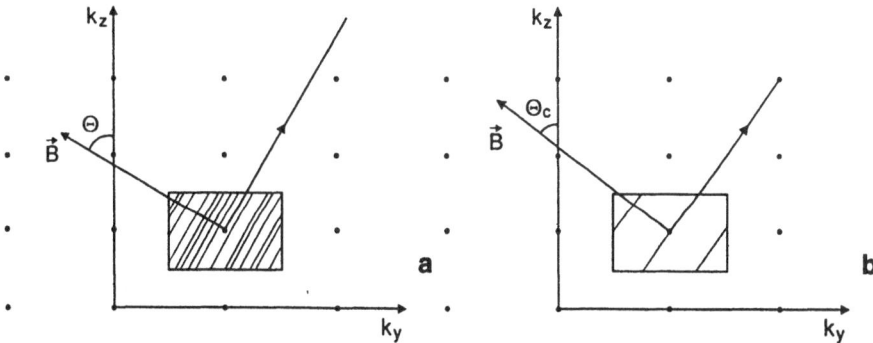

Fig. 3.5. Schematic view of an electron trajectory on the 1D FS sheet when the field B is applied (**a**) in an arbitrary direction in the bc plane and (**b**) in a commensurate direction $(p/q = 1/2)$

This commensurability effect and the magic angles were originally proposed as being responsible for a reduction of the FISDW transition temperature at these angles. For this field direction the electronic dispersion disappears along B [262]. Indeed, this effect has been observed experimentally in $(TMTSF)_2PF_6$ [263]. However, the further prediction [264] that in the normal state peaks in the magnetoresistance at the magic angles should appear is contrary to the observed resistance minima (see Fig. 3.4).

Some other approaches to explain the microscopic origin of the observed features have been suggested [265, 266]. In one model [266] the existence of so-called "hot spots" was proposed. These points on the FS sheets should have extremely high scattering rates for the electrons. The microscopic origin for these hot spots has not been given and is unclear. Furthermore, as was shown later [267] these spots cannot explain the magic angles and the angular dependence of the resistance.

Finally, in a recent proposal the experimental observations were explained without invoking any pecularities of the band structure [268]. The main idea of this theory is a field dependent renormalization of the coherent c-axis hopping, t_c, to zero. It was argued that single-particle hopping between spin-charge separated Luttinger liquids is incoherent for weak interliquid hopping

[269]. For magnetic fields of sufficient strength and pointing away from the magic-angle directions the material becomes essentially 2D by the vanishing of t_c. This 2D state is believed to be non-Fermi-liquid like [268]. If the field is applied along a real space lattice vector, $pb + qc$, higher order hops generated by the renormalization of t_b and t_c should be possible. This would explain the dips in the angular dependence of the resistivity.

In spite of the somewhat controversial theoretical picture the experimental signatures are unambiguous and can indeed be used to obtain useful informa-

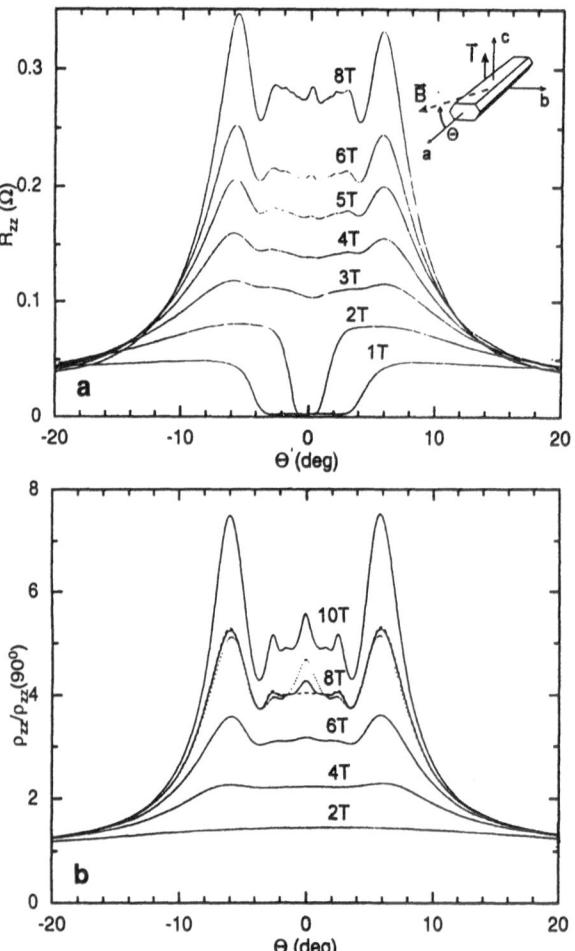

Fig. 3.6. (a) Angular dependence of the resistance in the c direction for $(TMTSF)_2$-ClO_4 at $T = 0.5\,K$ when the field is rotated in the ac plane (see inset). (b) Calculated AMRO using the formulas described in the text and the parameters $\tau = 4.3 \times 10^{-12}$ s, $t_b = 0.012\,eV$, and $t_c = t_b/15$ (solid lines). For 8 T results with $t_c = t_b/7.5$ (dashed line) and $t_c = t_b/60$ (dotted line) are shown. From [271]

tion regarding FS topologies. Especially for ET materials with combined 1D and 2D FS sheets resistivity minima at magic angles unravel the orientation of the 1D bands. As can be seen from (3.19), the physical information that one extracts from the AMRO effect is primarily related to the geometrical crystal structure, i.e., to the generally known lattice parameters. Further information that can be gained by systematic studies of the AMRO for different rotation planes is the exact orientation of the 1D planes with respect to the crystal axes. In the 2D materials, namely in α-$(ET)_2 M Hg(SCN)_4$, the observation of Lebed oscillations shown in Fig. 4.14 in the low-temperature DW state (see also Sect. 4.2.1) proved the existence of 1D FS sheets [270]. These 1D sheets, however, did not coincide with the predicted sheets of the band-structure calculation shown in Fig. 2.20. Therefore, a reconstruction of the FS with additional open sheets tilted by 24° due to this DW transition was proposed [270]. These measurements are discussed in more detail in Sect. 4.2.1.

Another AMRO effect was found in $(TMTSF)_2 ClO_4$ with the current flowing in the least-conducting c direction and when the field was tilted in the ac plane, in contrast to the bc plane discussed before. The experimental observations for fields close to the a direction are shown in Fig. 3.6a [271]. Above a certain field clear peaks in the magnetoresistance in the c direction are found. The explanation for this effect is also based on commensurability arguments [271]. The energy dispersion of the 1D materials close to ϵ_F can be approximated with a linear dispersion along x by (cf (2.4))

$$\epsilon(\boldsymbol{k}) = \hbar v_F(|k_x| - k_F) - 2t_b \cos(k_y b) - 2t_c \cos(k_z c). \tag{3.21}$$

One of the corresponding two FS sheets is shown schematically in Fig. 3.7 together with electron orbits for different field orientations. If the field is oriented parallel to the a direction a closed orbit 1′ and open orbits 1 exist. For field orientations slightly tilted away from a the trajectories have large amplitude oscillations along k_z and extend to infinity along k_y. For any arbitrary angle the average velocity in the z direction, v_z, might not be zero. However, if the oscillations in k_z just extend over an integer number of Brillouin zones (orbits 2 and 3) v_z averages to zero. Therefore, at these angles the resistivity in the z direction has local maxima. For angles larger than Θ for orbit 3, v_z can no longer average to zero. Therefore, the resistance in the z direction, R_z, decreases rapidly with increasing angle as shown in Fig. 3.6a.

The angles where the maxima in R_z occur are connected with the amplitude of the FS corrugation, that is $4t_b/\hbar v_F$, and the length of the Brillouin zone $K_z = 2\pi/c$. With the semi-classical equations (3.1), (3.2), and for $v_F B_z \gg v_z B_x$ the average velocity in the z direction is given by [271]

$$\langle v_z \rangle \propto \int_0^{2\pi} \sin[\gamma \cos(\Theta)] d\Theta, \tag{3.22}$$

with

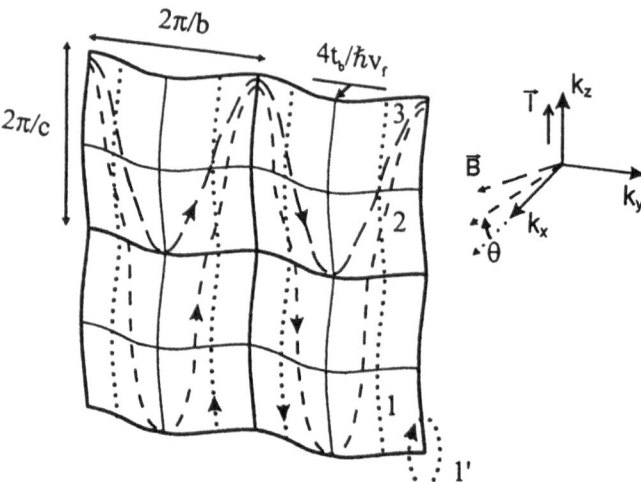

Fig. 3.7. Schematic FS of $(TMTSF)_2ClO_4$. Closed ($1'$) and open (1) orbits (dotted) for the field applied parallel to a. Paths 3 (long dashed) and 2 (short dashed) are the trajectories for angles where the field is near maxima of the resistivity data. From [271]

$$\gamma = \frac{2\, t_b c}{\hbar v_F} \frac{B_x}{B_z}. \tag{3.23}$$

The zeros of this integral are the same as for the Bessel function $J_0(\gamma)$. Therefore, from the positions of the resistivity maxima t_b can be extracted. The complete equation for the conductivity using the relaxation time approximation is given by [164]

$$\sigma_{ij} = e^2 \int \frac{d\mathbf{k}}{4\pi^3} \tau v_i(\mathbf{k})\bar{v}_j(\mathbf{k}) \left(-\frac{\partial f}{\partial \epsilon}\right), \tag{3.24}$$

where f is the Fermi function, τ is the constant scattering time, and

$$\bar{v}_j(\mathbf{k}) = \int_{-\infty}^{0} \frac{dt}{\tau} \exp(t/\tau) v_j(\mathbf{k}(t)). \tag{3.25}$$

By a numerical integration the result shown in Fig. 3.6b is obtained. Numerical and experimental results agree qualitatively. However, the quantitative magnetic field dependence of the oscillatory effect comes out somewhat too strong. Consistent with previous more indirect estimates $t_b = 0.012\,\mathrm{eV}$ is extracted. For the $8\,\mathrm{T}$ curve different values of t_c have been used. The fine structure of the resistivity data near $\Theta = 0$ depends on this parameter. The result of $t_c \approx t_b/15$ is in good agreement with the experimental data and previous estimates [28]. This latter value, however, can be extracted from the data only with large error. Nevertheless, this AMRO effect is a new method with which quantitative information on t_b, the transfer integral in the second least conducting direction for 1D materials, can be obtained.

4. The Fermi Surfaces

As was briefly discussed in Chap. 2, the principal Fermi surface topologies of 1D and 2D materials are simple and easily understood. The band structure in real low-dimensional organic metals, however, reveals some unusual and surprising pecularities. The exact knowledge of the FS is important in understanding the electronic and superconducting properties of metals. Measurements of possible changes of the FS under the influence of external parameters like pressure, magnetic field, and temperature give valuable information on the subtle balance of the competing interactions in organic substances. The question as to whether a lowering of dimensionality is advantageous for superconductivity or not, which parameters are responsible for the formation of SDW and CDW ground states, and how complete our basic understanding of low-dimensional materials is, can perfectly be addressed by FS studies in organic metals. The advantage of the organic materials as compared, for instance, with high-T_c cuprates is the availability of high-purity crystals with low scattering rates, τ^{-1}. The continuing progress in crystal growth techniques enables Fermi surface studies for these comparatively simple electronic systems in a fairly easy way, that is with standard available field and temperature ranges.

The band structures of a variety of different organic metals have been investigated since 1988. The principal FS topologies have been mapped out in detail. However, for the largest number of synthetic metals the FS has not yet been verified experimentally. For some other substances the published results are partially in contrast to each other or in disagreement with the calculated band structures.

In the following the experimental knowledge of the FS of the 1D Bechgaard salts is first summarized. Part of the results has already been mentioned in previous chapters. In the second part dHvA, SdH, and AMRO results for the 2D ET compounds organized with respect to the different phases are presented. Finally, some experiments for non-ET materials will be reviewed.

4.1 One-Dimensional Organic Metals

In the history of investigating 1D organic metals the discovery of SdH oscillations with frequencies reflecting small closed 2D or 3D orbits has been

claimed several times. This apparent contradiction to the 1D electronic nature was assumed to be an indication of the occurrence of a low-temperature
density wave (DW) state where parts of the FS sheets do not fit completely
on top of each other. FS parts which can be brought on top of each other by
the DW nesting vector vanish, whereas the other parts may remain as closed
areas which might be detectable by magnetic quantum oscillations.

However, these very low-frequency oscillations periodic in $1/B$ could be
successfully explained by the standard model for field induced spin density
waves [75, 76], see also Figs. 2.8 and 2.9 in Sect. 2.2.3. Although this effect
is strongly related to the quasi-1D band structure and is by itself highly
interesting no direct conclusions concerning the FS topology can be extracted
from the measurements.

The second effect, the rapid or fast oscillations, also briefly mentioned in
Sect. 2.2.3, shows many similarities to the usual SdH effect and was, therefore,
in the beginning interpreted as being due to closed orbits in the DW state.
However, as pointed out already, the resistance oscillations, although periodic
in $1/B$, show a temperature and magnetic field dependence which is not
understandable within the usual Lifshitz–Kosevich theory (3.6).

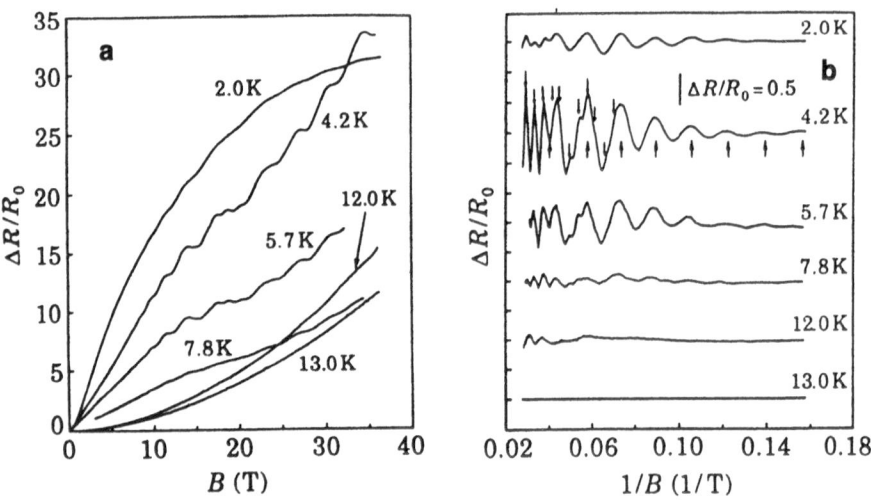

Fig. 4.1. (a) Relative magnetoresistance normalized to the zero field values of
$(TMTSF)_2NO_3$ for different temperatures. (b) Oscillatory part of the curves in (a).
Thick and thin arrows correspond to frequencies of 63 T and 248 T, respectively.
From [92]

As an example of rapid oscillations Fig. 4.1 shows the magnetic field dependence of the relative resistivity of $(TMTSF)_2NO_3$ for different temperatures [92]. The periodicity of two sets of oscillations in $1/B$ is clearly visible.
However, the oscillations are largest for 4.2 K and are reduced both for higher
and lower temperatures. A similar inconsistency with the predictions of (3.6)

can be seen in the field dependence of the oscillations which does not follow the behavior expected from the Dingle reduction factor (3.10).

A possible explanation of the experimental observations proposes that the oscillations are due to quantum interference effects in connection with magnetic breakthrough (MB) when the SDW gap is opening [92, 93, 272]. The magnetic interference effect is known in principle from pure Mg [273] where interference occurs between carriers traveling along open orbits which cross or come close to each other in at least two MB junctions. MB, on the other hand, occurs when the energy gap ϵ_g between two bands at some point in k space is small enough. The probability for an electron to go from one band to the other is given by the Blount criterion [274]

$$p = \exp\left(-\frac{B_{MB}}{B}\right),\qquad(4.1)$$

where

$$B_{MB} = \frac{\pi m_c}{4\,e\hbar\sin 2\,\Theta_{MB}}\frac{\epsilon_g^2}{\epsilon_F}\qquad(4.2)$$

is the MB field and Θ_{MB} is the Bragg reflection angle. Therefore, the MB condition occurs at a much smaller field than might be expected, viz ϵ_g^2/ϵ_F instead of ϵ_g. In the present case ϵ_g is assumed to be the temperature dependent SDW gap. The oscillating part of the interfering current is given by $I_{osc} = -2p(1-p)$. Therefore, this model would explain the occurrence of the largest oscillations at the temperature and field range when the gap is just opening, that is when p is neither 0 or 1. The frequency of the oscillations would be connected to the phase difference between the two current paths which is proportional to $1/B$ and the enclosed loop area.[1] The origin of these oscillations, however, is basically different from the SdH effect: the closed-loop integral of the generalized momentum (see (3.4)) does not need to be quantized. Although the proposed mechanism explains the principal features of the experimental observations a quantitative theoretical description and conclusive verification are still missing.

Angular dependent magnetoresistance measurements are to date the only possibility of deriving real topological information of the FS in 1D organic metals. As was shown in detail in Sect. 3.3, the measurement of the magnetoresistance for fields rotating in the ac plane can give the amplitude of warping in the b direction directly, that is the transfer integral t_b [271]. In addition, from the fine structure of the resistivity data close to $\Theta = 0$, i.e., for the field almost along the a direction, an estimate of t_c can be obtained.

The Lebed oscillations (or magic angle effect), on the other hand, are observed in materials with 1D FS sheets as AMRO for rotations of the field perpendicular to the highly conducting direction. These oscillations, however,

---

[1] The analogous oscillations in real space that are periodic in B are called the Aharonov–Bohm effect [275].

measure lattice parameters entirely. As long as no quantitative theory is available there is no possibility extracting any band-related parameters.

In conclusion, in 1D materials the topological aspects of the FS are clear in principle. However, an experimental technique to map out the FS in detail is just developing. In addition, theoretical understanding of the observed features in transport measurements is far from being complete in contrast to the well elaborated Lifshitz–Kosevich theory for 3D metals. Nevertheless, the principal simple quasi-1D form of the FS allows for an enormously rich variety of experimentally observable physical phenomena which were discussed partially in Sect. 2.2.3. These have stimulated a large amount of theoretical effort to obtain a better understanding of these effects in relation to the electronic band structure.

4.2 Two-Dimensional Organic Metals

The experimental picture dealing with the FS of 2D organic metals is considerably more elaborate than that of the 1D materials. Meanwhile, a variety of different SdH, dHvA, and AMRO data exist which have helped to elucidate the electronic structure of the 2D compounds. Since the band structure naturally depends strongly on the crystal structure the different phases of the ET materials are discussed separately in the next sections. Experimental results of non-ET compounds are reviewed in the last part of this section.

4.2.1 α Phase

The principal crystal structure of the so-called α phase of $(ET)_2X$ shown in Fig. 2.16 and some superconducting properties had already been known shortly after the first observation of ET-based superconductors [145]. More especially, the discovery of the annealed phase α_t-$(ET)_2I_3$ with a drastically increased $T_c \approx 8\,K$ initiated interest in this phase [276]. In these investigations it was found that annealing α_t-$(ET)_2I_3$ above $70\,°C$ for several days transformed this phase into a β_H-related phase of $(ET)_2I_3$ (see Sect. 2.3) which is stable under ambient pressure and room temperature.

The FS investigations, however, started only after the discovery of the α-phase family $(ET)_2MHg(YCN)_4$ with $M = K$, NH_4, Rb, Tl and $Y = S$ or Se. The large number of investigations of these compounds is stimulated by the subtle balance between a low-temperature density-wave state and normal metallic and superconducting behavior which can be tuned comparatively easily in the $P-B-T$ parameter space. The coexistence of closed 2D and open 1D FS sheets (see Fig. 2.20) gives the possibility investigating the influence of different dimensionalities within the same material.

The only superconductor of this family is α-$(ET)_2NH_4Hg(SCN)_4$ with $T_c \approx 1\,K$ [146]. The sulfur-based salts with $M = K$, Rb, and Tl may show some superconducting traces [277] but do not show bulk superconductivity down to ~50 mK. These latter materials show some kind of phase transition between 8 and 12 K which seems to be connected with a FS reconstruction due to a SDW transition. An "intermediate" role is played by α-(ET)$_2$TlHg(SeCN)$_4$ and α-(ET)$_2$KHg(SeCN)$_4$ which neither show superconductivity nor a SDW transition.

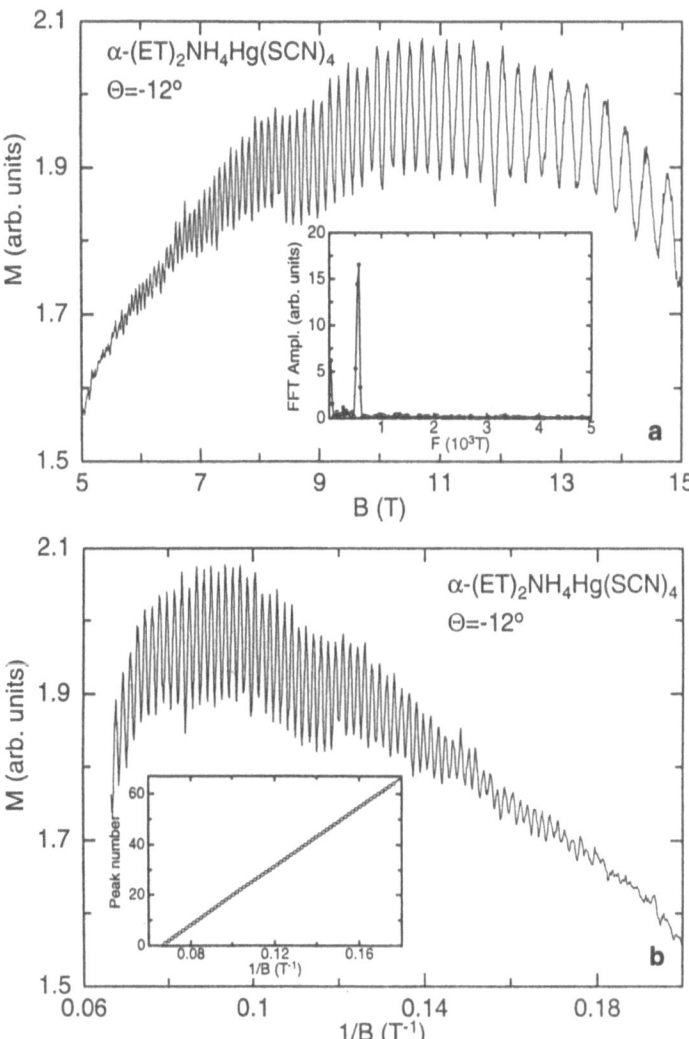

Fig. 4.2. dHvA signal of α-(ET)$_2$NH$_4$Hg(SCN)$_4$ at $T = 0.5$ K vs (a) magnetic field and (b) inverse magnetic field, applied at an angle of $\Theta = -12°$ normal to the conducting planes. Inset (a) shows the FFT of the data between 10 and 15 T. Inset (b) shows the arbitrary peak number vs inverse field

Meanwhile, an enormous number of SdH and dHvA investigations of the α-phase family exist. For the superconductor with $M = \mathrm{NH_4}$ the picture obtained is quite conclusive and will be presented first. Figure 4.2 shows typical results of a comprehensive dHvA investigation revealing the FS topology, the angular dependence of the effective masses, and many-body renormalization effects [278, 279]. For the measuring temperature of $T \approx 0.5\,\mathrm{K}$ and for a fixed field direction (the angle between field B and the normal to the ET planes is $\Theta = -12°$) clear dHvA oscillations with only one frequency are visible starting at $\sim 5.5\,\mathrm{T}$. From the plot of M vs the inverse field in Fig. 4.2b the periodicity of the signal in $1/B$ can be seen. The inset in (a) shows the fast Fourier transformation (FFT) of the data between 10 and 15 T. The FFT reveals the spectral purity of the signal with almost no harmonics at this angle. For the experimentally available field range the determination of the exact dHvA frequency, F, from the FFT, however, is limited to an accuracy of approximately $\pm 20\,\mathrm{T}$. F can be extracted in a more accurate way ($\pm 1\,\mathrm{T}$) from the linear regression of a plot of the peak number vs $1/B$ shown in the inset of Fig. 4.2b. For high fields a slight reduction of the dHvA signal can be seen. This is due to the modulation field used in the present experiment. As the modulation-field amplitude was kept constant, therefore, the signal is reducing for large B which can be quantitatively analyzed with (3.17).

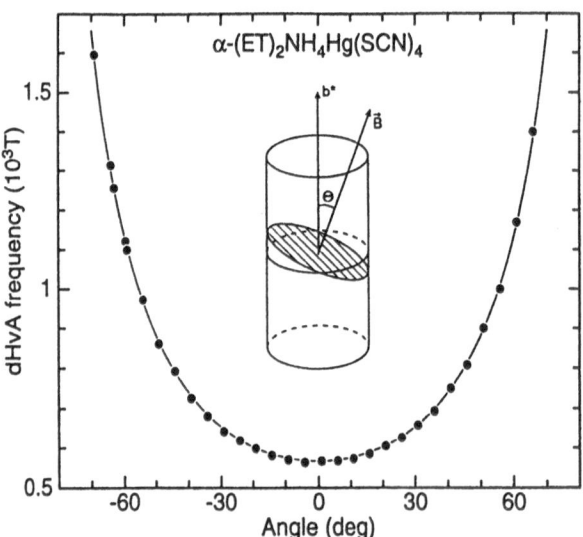

Fig. 4.3. Angular dependence of the dHvA frequency in α-$(\mathrm{ET})_2\mathrm{NH_4Hg(SCN)_4}$ with the solid line showing the $1/\cos\Theta$ behavior. The inset depicts the relative orientation of the field with respect to the 2D FS

The angular dependence of the oscillation signal has to be measured to obtain a more complete picture of the FS. For α-$(\mathrm{ET})_2\mathrm{NH_4Hg(SCN)_4}$ it was possible to observe dHvA oscillation up to angles above 70°. The angular dependence of the dHvA frequency is shown in Fig. 4.3 [278, 279]. F follows the behavior expected for an ideal 2D cylindrical FS perfectly, namely $F = F_0/\cos\Theta$, which is plotted as the solid line in Fig. 4.3. This functional dependence has also been found in most of the reported angular dependent SdH and dHvA measurements of the 2D ET materials.[2] The minimum frequency $F_0 = (566.7 \pm 1)$ T is in agreement with SdH experiments [280] and corresponds approximately to the closed area around the V point of the Brillouin zone predicted by band-structure calculations (see Fig. 2.20) [141].

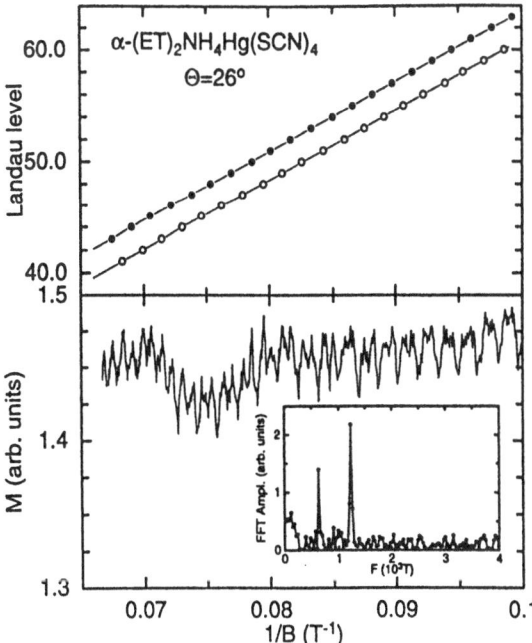

Fig. 4.4. The lower part shows the magnetization of α-(ET)$_2$NH$_4$Hg(SCN)$_4$ vs reciprocal field at $\Theta = 26°$ and $T = 0.5$ K. The inset shows the FFT of the data. The upper part shows the corresponding Landau level number for levels with spin parallel and spin antiparallel to B. The solid lines are linear fits yielding a dHvA frequency $F = (627 \pm 1)$ T

While the dHvA frequency follows the expected behavior of 2D electronic systems smoothly the angular dependence of the dHvA amplitude reveals drastic and at first sight unexpected effects. For $\Theta = 26°$ the dHvA signal shown in Fig. 4.4 was observed. The same principal behavior was found for $\Theta = -26°$ [279]. Compared to the signal shown in Fig. 4.2 the amplitude of the signal is reduced by a factor of ~ 8 and shows a double-peak-like structure.

---

[2] Note that even if the FS cylinder is corrugated the average frequency F_{av} follows the $1/\cos\Theta$ dependence, see section 4.2.2.

The FFT shown in the inset clearly reveals two peaks. The peak at $F_1 \approx 630$ T corresponds to the expected cross-sectional area of the FS cylinder from the $1/\cos\Theta$ dependence. The even larger second peak is exactly at $2F_1$ and is the second harmonic of the dHvA signal. Obviously, at this angle the spin-splitting factor $R_S = \cos(r\pi g\mu_b/2)$ described in Sect. 3.1 is close to zero for the fundamental ($r = 1$) and maximum for the second harmonic ($r = 2$).

In order to gain more information about $g\mu_b$ the number of the two sets of Landau levels with spin up and spin down are plotted vs the corresponding inverse field in the upper part of Fig. 4.4. A simultaneous linear regression of both data sets results in an average dHvA frequency $F = (627 \pm 1)$ T. The field B where the nth Landau level just crosses the FS is given by [249]

$$F/B = n + \gamma \pm \frac{1}{2}S, \tag{4.3}$$

where the spin-splitting parameter is defined as $S = (1/2)g\mu_b$. The difference between the fit-line intercepts at $F/B = 0$ gives $S = 2.43 \pm 0.1$, γ is determined to 0.54 ± 0.1 in accordance with the expected value of ~ 0.5. Of course, without further information the stated value of S is arbitrary and the assigned numbers of the Landau levels ambiguous. However, as will be shown later, the value of $g\mu_b \approx 5$ at this angle is the most plausible one.

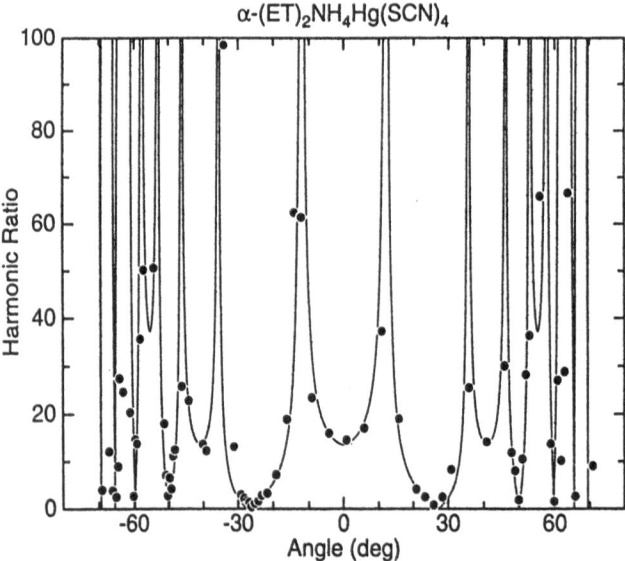

Fig. 4.5. Angular dependence of the harmonic ratio of the measured dHvA oscillations for α-(ET)$_2$NH$_4$Hg(SCN)$_4$. The solid line is calculated using (4.4)

Systematic investigations of the angular dependence of the dHvA amplitude in α-(ET)$_2$NH$_4$Hg(SCN)$_4$ showed at least three additional spin-splitting zeros. Between these, the fundamental amplitude has maxima. This behavior can more clearly be seen in Fig. 4.5 where the measured harmonic ratio (HR), i. e., the ratio of the fundamental $(r = 1)$ dHvA amplitude, M_1, to the second harmonic $(r = 2)$, M_2, is plotted vs angle. For not observable second harmonics (as in Fig. 4.2 at $\Theta = -12°$) and for angles greater than $\pm 50°$ the noise level of the FFT was taken as an upper limit for M_2. Therefore, the data points for large HR in Fig. 4.5 are somewhat underestimated.

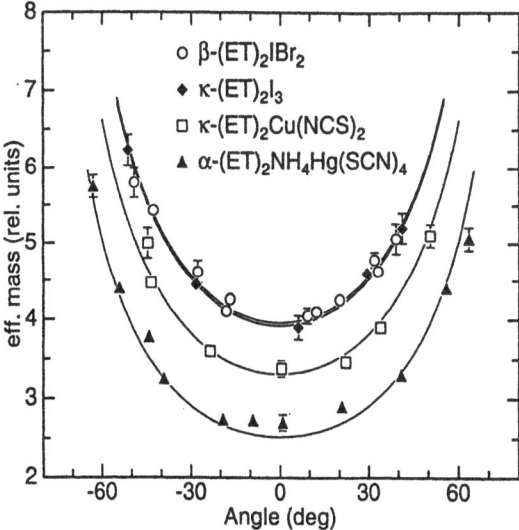

Fig. 4.6. Angular dependence of the effective cyclotron mass of four organic superconductors. The solid lines are of the form $m_c = m_{c0}/\cos\Theta$

In order to understand the unusual behavior of the HR one has to know the relevant parameters. From the standard Lifshitz–Kosevich formula (3.6) the HR can easily be derived [249, 281] and is given by

$$\frac{M_1(\Theta)}{M_2(\Theta)} = \left|\frac{\cos(\pi g\mu_b/2)}{\cos(\pi g\mu_b)}\right| \frac{2\sqrt{2}\cosh(\alpha\mu_c T/B)}{\exp(-\alpha\mu_c T_D/B)} J_F, \qquad (4.4)$$

where J_F is a factor which for the modulation field method is given by the Bessel function ratio

$$J_F = \frac{J_n(2\pi F b_0/B^2)}{J_n(4\pi F b_0/B^2)}, \qquad (4.5)$$

see also (3.17). Obviously, $g\mu_b$ must be strongly angle dependent to account for the observed HR behavior. The effective mass averaged over a cyclotron orbit as given by (3.3) is proportional to the derivative of the extremal FS area with respect to energy. Therefore, since the area changes with $1/\cos\Theta$ this dependence should for the first approximation be also reflected in the effective mass. Systematic investigations of the angular dependence of the effective cyclotron mass did indeed reveal a $1/\cos\Theta$ dependence. Figure 4.6 shows m_c determined from the temperature dependence of the dHvA amplitude for four different ET based superconductors (see also Fig. 3.1). The data can be fitted comfortably within the function $m_c = m_{c0}/\cos\Theta$ shown by the solid lines with values m_{c0} of $(2.6 \pm 0.1)\,m_e$ for α-$(ET)_2NH_4Hg(SCN)_4$, $(3.3 \pm 0.1)\,m_e$ for κ-$(ET)_2Cu(NCS)_2$, $(3.9 \pm 0.1)\,m_e$ for κ-$(ET)_2I_3$, and $(4.0 \pm 0.1)\,m_e$ for β-$(ET)_2IBr_2$.

With the experimentally proven $1/\cos\Theta$ dependence of m_c the distinctive HR behavior is now easily understandable. The effective cyclotron mass without electron–phonon interaction, $\mu_b = m/m_e$, shows up in the argument of the spin-splitting factor $\cos(r\pi g\mu_b/2)$. Therefore, at angles where $g\mu_b = 2n - 1$ the fundamental dHvA amplitude becomes zero and for $g\mu_b = 2n - (1/2)$ and $g\mu_b = 2n + (1/2)$ the second harmonic is zero. From the six experimentally best-defined extremal points (averaged over both $(\pm)$ angular directions) of the HR (see Fig. 4.5) and with the assumption $\mu_b = \mu_{b0}/\cos\Theta$ the most probable values for $g\mu_{b0}$ are plotted in Fig. 4.7 vs angle. The data only increase slightly (by approximately 4%) with angle which within the accuracy can be approximated by a linear dependence. However, there is some indication for a stronger increase of $g\mu_{b0}$ close to $\Theta = 0°$. Whether this is due to an angular change of g or a deviation from the $m_c \propto 1/\cos\Theta$ dependence is not clear at the moment.

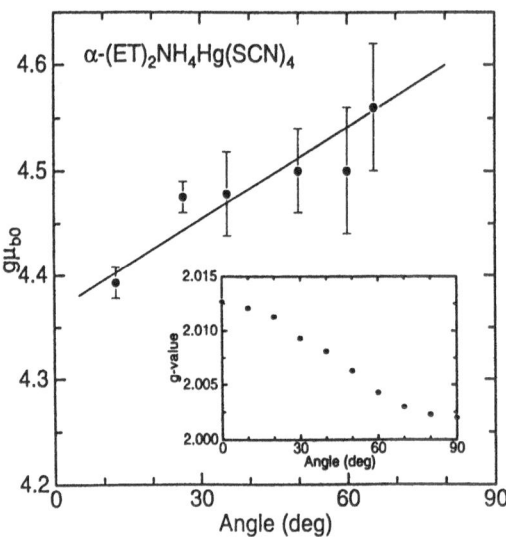

Fig. 4.7. Angular dependence of $g\mu_{b0}$. The solid line is a linear approximation to the data. From [278, 279]. The inset shows the angular dependence of the g value determined by ESR experiments. From [146]

The angular dependence of the g value of α-$(ET)_2NH_4Hg(SCN)_4$ obtained from ESR measurements [146] is shown in the inset of Fig. 4.7. This ESR value is nearly equal to 2 and decreases only very weakly (by approximately 0.5%) with Θ. With $g = 2$ one obtains the average value $\mu_{b0} \approx 2.23$. Together with the well-known other parameters ($T = 0.5\,K$, $B = 14\,T$, $T_D = 0.5\,K$), excellent quantitative agreement between data and the calculated HR using (4.4) is obtained. This is shown by the solid line in Fig. 4.5. This agreement means that the 3D Lifshitz–Kosevich formula for the investigated field and temperature region is still fully applicable for the 2D metal α-$(ET)_2NH_4Hg(SCN)_4$.

With the measured values of both effective masses the electron–phonon coupling constant $\lambda = \mu_c/\mu_b - 1 = 0.17$ can be extracted. Using the simple BCS formula (2.6), the superconducting transition temperature can be estimated. With the Debye temperature $\Theta_D = 230\,K$ known from specific-heat measurements [211] $T_c \approx 0.6\,K$ is obtained. The fact that this rough estimate is in the right range of the experimentally observed T_c of $\sim 1\,K$ might be somewhat accidental but it indicates that the existence of electron–phonon interaction in at least this organic superconductor is highly suggestive. The other possibility of a g value of only 1.73 which also would explain the low value of $g\mu_{b0}$ seems to be not very likely.

The experimental determination of both μ_c and μ_b from the same set of dHvA data is unusual but rendered possible because the effective mass changes strongly but is ascertainable with angle. In addition, the value of m_b is in an advantageous range of a few times the free electron mass. This causes the occurrence of a number of observable spin-splitting zeros necessary to obtain an unambiguous condition for $g\mu_b$. The extraction of μ_b with the ESR g-value of nearly 2 is straightforward but, of course, not devoid of problems. In ESR experiments the g value is not enhanced by electron–electron interactions whereas these have an influence on both dHvA masses, μ_c and μ_b. Since the electronic system is highly 2D and the density of carriers is rather low in the ET materials electron correlation effects might play an important role.

Some indications for strong Coulomb interaction effects have been obtained from cyclotron resonance studies [282] where a cyclotron resonance mass m_{CR} between 1.05 and $1.4\,m_e$ was found. This value is close to the effective mass extracted from the band-structure calculations. A value of $\sim 1\,m_e$ was predicted. The cyclotron resonance probes the energy separation between Landau levels at and above ϵ_F. The effective cyclotron mass m_c, on the other hand, occurs in the thermodynamic density of states and includes contributions from interacting quasiparticles. The corresponding renormalizations are only effective close to the Fermi energy. If the cyclotron energy is larger than the renormalized region m_{CR} might be quite different from m_c [283]. Therefore, it was assumed that very strong electron–electron correlations are present in organic superconductors [284]. In usual metals the mass enhancement due to Coulomb correlation effects is only a few percent.

This is drastically different for heavy Fermion compounds where extreme enhancement factors are known. Therefore, the organic superconductors were compared and related to this latter unique class of metals [284].

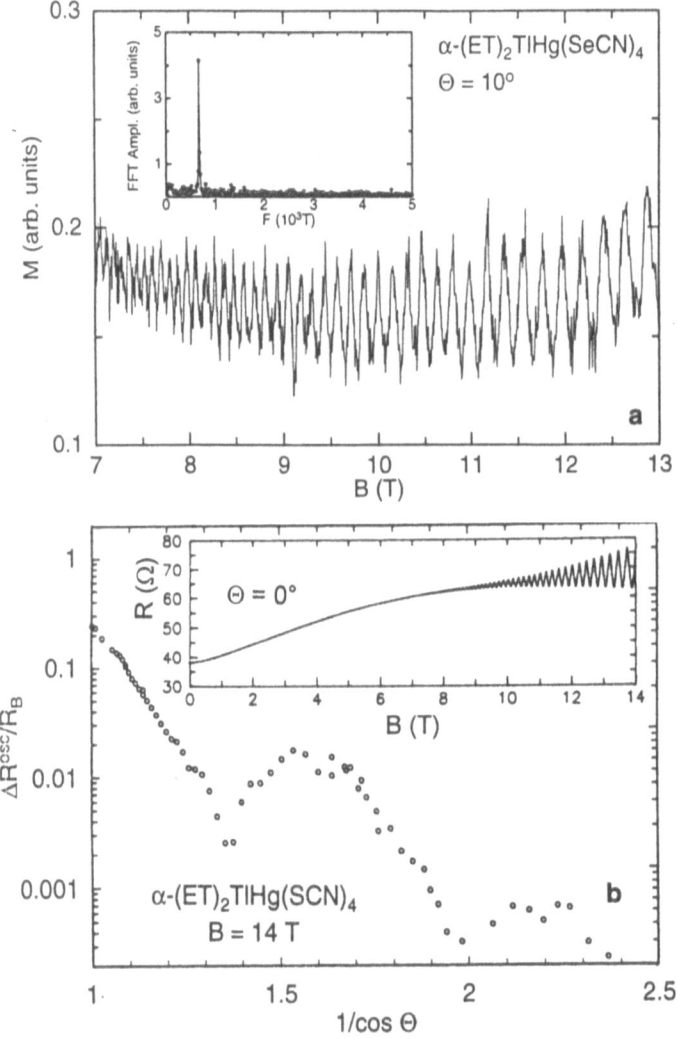

Fig. 4.8. (a) dHvA signal of α-(ET)$_2$TlHg(SeCN)$_4$ at $\Theta = 10°$ and $T \approx 0.5\,\mathrm{K}$. The inset shows the FFT of the data. (b) Angular dependence of the SdH amplitude. From [285]. The inset shows the SdH signal of the same sample at $\Theta = 0°$

The next α-phase salt to be discussed is α-(ET)$_2$TlHg(SeCN)$_4$. As already mentioned, this compound is not superconducting down to $\sim 80\,\mathrm{mK}$. The measurements of both SdH [285] and dHvA effect [286] revealed a similar picture as for α-(ET)$_2$NH$_4$Hg(SCN)$_4$. Typical dHvA oscillations of one

α-$(ET)_2TlHg(SeCN)_4$ sample are shown in Fig. 4.8a. Although the data are somewhat noisy because of the small sample size clearly one dHvA frequency of $F \approx 670\,T$ together with a hardly detectable second harmonic are found in the FFT. The inset of Fig. 4.8b shows SdH oscillations of the same sample [285]. Both experiments are fully consistent with each other in the extremal area, the effective mass, and the angular dependences of both. The measurements resulted in the typical 2D $1/\cos\Theta$ behavior of the FS area with $F_0 \approx 660\,T$. This value is considerably larger than in α-$(ET)_2NH_4Hg(SCN)_4$ although the unit cell in the latter salt is even slightly smaller, i.e., the Brillouin zone is slightly larger (see also Tab. 2.1). Since for free electrons in a 2D metal $\epsilon_F \propto n/m$, where n is the electron density, one would expect a smaller effective mass in the $X = TlHg(SeCN)_4$ salt. Indeed, the effective cyclotron mass determined from the temperature dependence of the oscillation amplitude is $m_c = (2.0 \pm 0.1)\,m_e$. This value, however, is roughly 30% smaller than m_c found in the $M = NH_4$ compound, whereas the area is only 16% larger. Therefore, some more complicated effects must play a role in these isostructural substances.

These experiments revealed detectable spin-splitting zeros of the SdH and dHvA oscillations for two angles. In Fig. 4.8b these angles are located at the points where the absolute amplitude of the SdH oscillations is minimal. In a recent dHvA experiment extending the field range up to 29 T more spin-splitting zeros could be determined with high accuracy [287]. From these zeros a value of $g\mu_b \approx 3.7$ can be extracted. For $g \approx 2$ this gives $\lambda \approx 0.08$ which is clearly less than the value found in α-$(ET)_2NH_4Hg(SCN)_4$ and is compatible with the absence of superconductivity (T_c would be of the order of 1 mK estimated with (2.6)). On the other hand, α-$(ET)_2TlHg(SeCN)_4$ seems to be just on the border becoming superconducting.

SdH measurements on a sample at very low temperatures (80 mK) and for fields up to $\sim 35\,T$ have revealed a highly anharmonic resistivity oscillations as shown in Fig. 4.9 [288]. Extremely large SdH amplitudes with $\Delta R/R$ up to 800% and a highly asymmetric wave form of the oscillations were observed. The extraction of the effective cyclotron mass from the temperature dependence of the oscillation amplitude at different fields resulted in an apparent increase of μ_c with field [288]. As was noted in detail in Sect. 3.2, the extraction of μ_c out of resistivity data (and not out of conductivity oscillations) might be invalid for the large SdH effect observed here. Possibly one is already close to an ideal 2D case where strong harmonics are expected and where the usual 3D Lifshitz–Kosevich formula (3.6) is no longer valid. Therefore, the determination of μ_c with this formula must yield wrong results, especially when extracted from resistivity oscillations. Indeed, recent dHvA measurements up to 29 T have shown neither an unusual harmonic content of the oscillating signal nor any measurable increase of the effective cyclotron mass [287]. Additional further investigations of both SdH and dHvA effects,

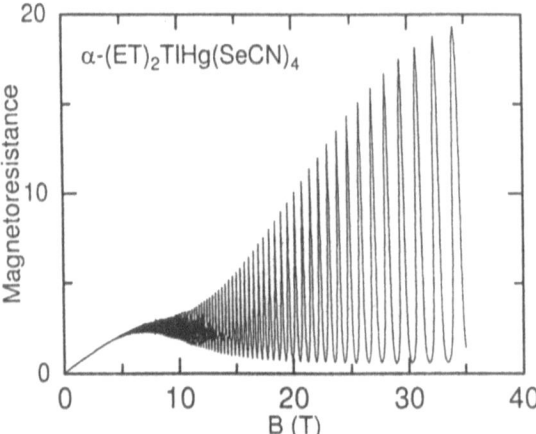

Fig. 4.9. Magnetoresistance of α-(ET)$_2$TlHg(SeCN)$_4$ at $T = 80\,\mathrm{mK}$ and up to 35 T. An extremely large amplitude and anharmonic wave form is observed. From [288]

preferably on the same crystal, would be highly desirable to elucidate these somewhat controversial results.

Recently another member of the α-phase family has successfully been synthesized [144]. α-(ET)$_2$KHg(SeCN)$_4$ behaves in many respects similar to α-(ET)$_2$TlHg(SeCN)$_4$. It neither becomes superconducting nor does it show a ground state instability, i. e., no resistivity hump around 10 K is observed. The band-structure parameters obtained from SdH measurements are almost the same for both materials. The SdH frequency follows the $1/\cos\Theta$ dependence with $F_0 \approx 670\,\mathrm{T}$. The effective cyclotron mass is $\mu_c \approx 2.0$. No mass increase up to 23 T was observed [144].

The other salts of the α-(ET)$_2M$Hg(SCN)$_4$ family with $M = $ K, Rb, and Tl all are not superconducting but show a hump in resistivity around $T_A = 8 - 10\,\mathrm{K}$ which is believed to be an indication for a SDW transition. As discussed already in Sect. 2.2.3, this is connected with a FS reconstruction. Consequently, magnetic quantum oscillations and AMRO measurements revealed an unusual and distinctive behavior. Although the details of the reconstructed FS in the SDW state are still discussed controversially the experiments helped to uncover some aspects of the FS topologies in the different states and showed that indeed the band structure is reconstructed in the low-field low-temperature state. In addition, at least some suggestions as to why these charge transfer salts are not superconducting were obtained.

The first reported SdH oscillations of α-(ET)$_2$KHg(SCN)$_4$ are shown in Fig. 4.10 [289]. Only one SdH frequency of approximately 667 T is observed. This value is the same as for α-(ET)$_2$TlHg(SeCN)$_4$ and α-(ET)$_2$KHg(SeCN)$_4$ within the error. The effective mass, however, was extracted to be $m_c = 1.4\,m_e$ which is considerably smaller than in the salts discussed before. In later

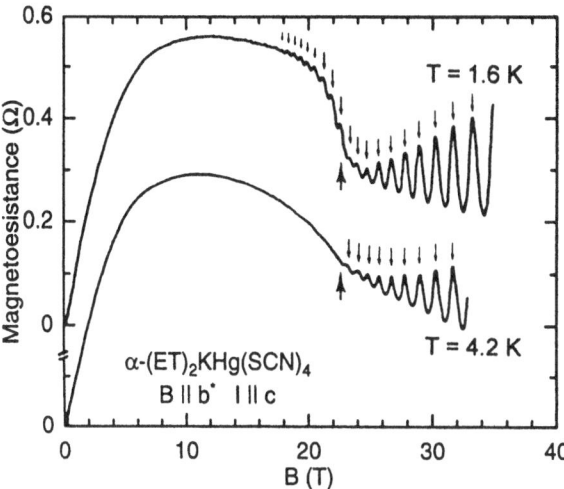

Fig. 4.10. Magnetoresistance of α-(ET)$_2$KHg(SCN)$_4$ for B applied perpendicular to the conducting plane. The large arrows indicate the kink field. The small arrows mark the SdH oscillation peaks. From [289]

experiments of one other group an effective cyclotron mass of $(1.9 \pm 0.1)\,m_e$ was reported [290]. The authors explain this difference by the insufficient subtraction of the background magnetoresistance of other groups. On the other hand, the determination of m_c from dHvA data by the same group [291] as well as another group [292] gave values fully consistent with the former results. Therefore, a cyclotron mass of $1.4\,m_e$ seems to be the most probable.[3]

The other important and unusual feature observed in the magnetoresistance is the strange behavior of the background resistance. As can be seen in Fig. 4.10, first the resistance increases drastically in the applied magnetic field and then falls off gradually above $\sim 10\,\mathrm{T}$. At approximately $23\,\mathrm{T}$ a sharp "kink" structure occurs after which the oscillation amplitude starts to increase rapidly. The kink structure in resistivity has been mapped out in detail by different groups [294, 295] and is believed to be the indication for the field-induced phase transition from the SDW to the metallic state.

Figure 4.11 shows the $B - T$ phase diagram of α-(ET)$_2$KHg(SCN)$_4$ [294]. The exact nature of the low-field low-temperature state is not known. Antiferromagnetic ordering has been suggested in view of the anisotropic static magnetic susceptibility [296] and the anomalous behavior of the electron spin resonance [297] below T_A. To date, however, only one μSR experiment [168] was able to verify the existence of a very small magnetic moment in the low-temperature state.

---

[3] Recently new dHvA results seem to indicate a field dependent decrease of m_c from $\sim 1.6\,m_e$ at $8\,\mathrm{T}$ down to $\sim 1.35\,m_e$ at $12.5\,\mathrm{T}$ [293].

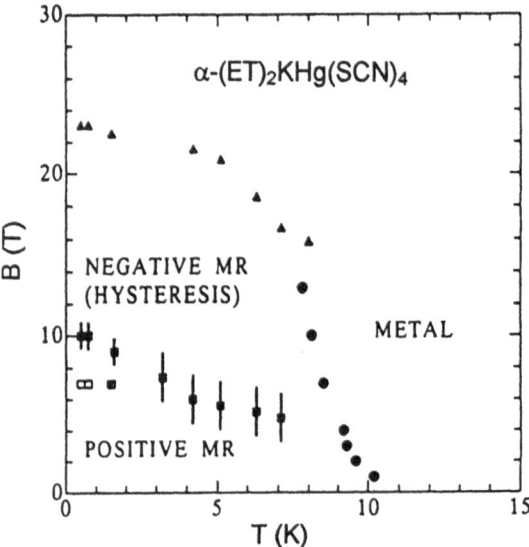

Fig. 4.11. Magnetic phase diagram of α-(ET)$_2$KHg(SCN)$_4$ obtained from magnetoresistance (MR) and magnetization anomalies. From [294]

A closer examination of the SdH and also dHvA oscillations with better-quality samples, i. e., samples with smaller Dingle temperature, revealed an unexpected splitting of the oscillation signal in the low-field region [298, 299]. Above the kink field, normal unsplit oscillations are observed. The effective cyclotron mass at these high fields was found to be $2.7\,m_e$ [291] which is close to the value found for the superconductor α-(ET)$_2$NH$_4$Hg(SCN)$_4$. The Dingle temperature was reduced appreciably [291]. A typical dHvA and SdH result up to 27 T is shown in Fig. 4.12 [295]. The oscillation shape is reminiscent of the spin-splitting effect in α-(ET)$_2$NH$_4$Hg(SCN)$_4$ shown in Fig. 4.4. Indeed, the first experiments were interpreted in this way. With the effective cyclotron mass of $\sim 1.4\,m_e$ [289, 298, 299] and $g \approx 2$ from ESR measurements [297] a spin-splitting parameter $S = 1.4$ is obtained neglecting a possible mass enhancement by electron–phonon coupling.

Later experiments revealed that the double-peak structure in α-(ET)$_2$-KHg(SCN)$_4$ must be of different origin. First, the effect is largely independent of the angle of the applied field and, second, the relative positions of the two peaks are shifting in field with respect to each other [290, 294, 295, 296, 300, 301]. Different approaches have tried to explain this shift. Due to the antiferromagnetic order in a SDW state it was assumed that magnetic interaction with the intrinsic magnetic moments might play a role in α-(ET)$_2$KHg(SCN)$_4$ [302]. The experimental data could be described very well by a field-dependent g factor $g = g_\infty + B_{ex}/B$, where the exchange field $B_{ex} = \Delta E/\mu_B$ and g_∞ are fit parameters [290, 302]. Another suggestion proposes a mixture of SDW and the metallic state each with a slightly differ-

ent extremal FS area in the samples [303]. This would partially explain the
sample and cooling rate dependence found by different groups. One group
[290, 300], for example, has found other frequency components at 150–200 T
and 820–870 T with amplitudes and frequencies being sample and cooling dependent in addition to the fundamental frequency of ~ 670 T. These effects
were ascribed to different MB orbits which should become relevant in the
SDW state. The change in frequency from run to run was proposed to be due
to a partially random nesting vector. In summary, in spite of the large number of different investigations the observed spin-splitting features of the SdH
and dHvA effects are not well understood in this material and necessitate
further studies.

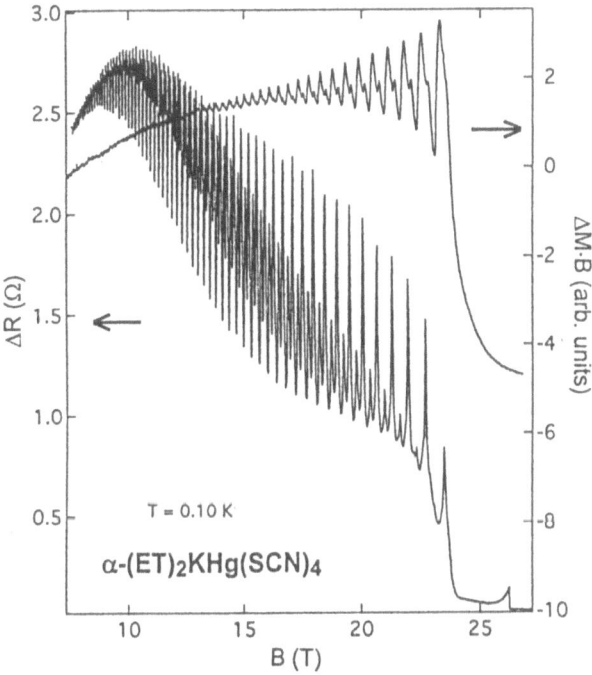

Fig. 4.12. Field dependence of the magnetoresistance and magnetization of
α-$(ET)_2KHg(SCN)_4$ showing the doublepeak structure of the oscillations. From
[295]

To complicate this picture even more, SdH experiments at very low temperatures in the SDW state revealed an additional high-frequency oscillation
at $F_{high} \approx 4260$ T as shown in Fig. 4.13a [304]. Recent dHvA measurements
down to 20 mK verified this finding [293]. The corresponding orbit was assumed to be due to MB in the unmodified metallic FS as shown schematically
in Fig. 4.13b. This suggestion, however, is highly speculative since the energy

gap between the 2D and 1D band obtained from band-structure calculations [141] (see also Fig. 2.20) is much too large to expect an observable MB effect for the experimentally reached field range. In addition, the FS in the SDW state should be largely reconstructed. Therefore, the proposed MB orbit can only exist if either the SDW does not reconstruct the FS or there does indeed exist a mixture of SDW and the metallic state [303].

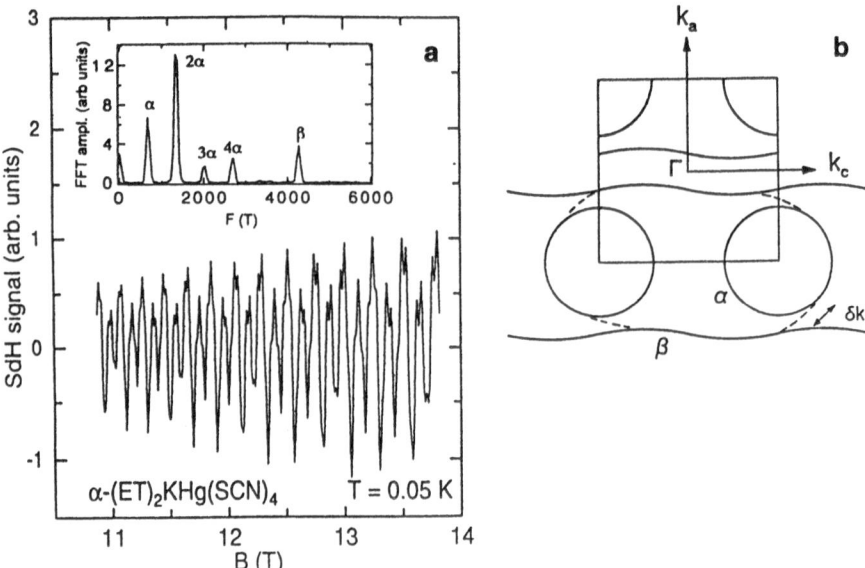

Fig. 4.13. (a) SdH signal of α-(ET)$_2$KHg(SCN)$_4$ for $\Theta = 0°$ at $T = 50$ mK. The FFT shown in the inset reveals in addition to the usual fundamental double-peak, α and 2α, with harmonics a new frequency β at 4260 T. (b) Proposed orbits in the 2D FS. The β orbit should occur between the usually seen closed 2D orbits and the 1D FS sheets due to magnetic breakdown. From [304]

For the other sulfur salt α-(ET)$_2$TlHg(SCN)$_4$ the same general observations as just discussed for α-(ET)$_2$KHg(SCN)$_4$ have been made. First experiments revealed a double-peak SdH oscillation of ~ 670 T and an effective mass of $\sim 1.4\,m_e$ exactly as in the latter salt [305]. Later, in addition to this low frequency a (MB) high frequency of 4270 T was observed [306]. Further on, in this salt three SdH-like oscillations with very low frequencies below 13 T have been reported [306]. These orbits were ascribed to small remaining parts of the reconstructed FS in the SDW state, whereas the fast oscillation was again explained by the MB orbit in the unreconstructed FS.

The last compound of this family, α-(ET)$_2$RbHg(SCN)$_4$, also reveals a behavior remarkably similar to that of α-(ET)$_2$KHg(SCN)$_4$. The SdH oscillations show a double-peak like structure with a frequency of ~ 654 T [307].

From the sharp structures in the magnetoresistance the kink fields and the corresponding phase diagram were mapped out. The latter was found to be very similar to the one shown in Fig. 4.11 for α-(ET)$_2$KHg(SCN)$_4$; however, with a somewhat higher value of ~ 30 T at $T = 1.2$ K [308].

That a reconstruction of the FS occurs below the ordering temperature T_A defined by the kink field has more directly been verified by AMRO experiments. When the magnetic field was rotated in planes perpendicular to the highly conducting crystal plane oscillations in the resistance not caused by the usual SdH effect could be observed in a variety of different 2D ET materials. As mentioned in Sect. 3.3, maxima periodic in $\tan\Theta$ can be related to a corrugated 2D FS as shown for β-(ET)$_2$IBr$_2$ in Fig. 3.2 [172]. The same effect has been reported for α-(ET)$_2$NH$_4$Hg(SCN)$_4$ [280] and α-(ET)$_2$TlHg(SeCN)$_4$ [285] reflecting the warping of the 2D FS in these materials. This warping, however, must be very small, less than 0.5% of the FS area. Otherwise a beating pattern of the dHvA and SdH oscillations (as will be discussed later) would have been observed.

All other α-phase salts (M = K, Rb, and Tl) showed AMRO periodic in $\tan\Theta$, too. In contrast to the magnetoresistance shape in the former materials no resistance maxima as expected for a 2D warped cylinder but instead sharp resistance minima were observed [289, 309, 310, 311]. Figure 4.14a shows examples of this behavior for α-(ET)$_2$KHg(SCN)$_4$ and α-(ET)$_2$RbHg(SCN)$_4$ at $B = 11$ T and $T = 1.4$ K [312]. Very prominent minima periodic in $\tan\Theta$ are visible in the magnetoresistance. These minima cannot be explained by the 2D warped FS model but are attributed to Lebed-like oscillations of the 1D bands [313] (see also Sect. 3.3). From the calculated band structure shown in Fig. 2.20 one can see that the FS of the α phase consists of 2D closed orbits in the Brillouin zone corner and two 1D open sheets perpendicular to the a direction. The maximum amplitude of the AMRO, however, was not observed for a rotation of the field parallel to these sheets but for angles 20° and 24° towards the a direction of the M = K and M = Rb crystals, respectively (see Fig. 4.14a).

The tilt angle Φ_0 can be determined more accurately from a plot of the oscillation period $\Delta\tan\Theta$ vs Φ as shown for the M = Rb salt in Fig. 4.14b. The Lebed-like AMRO period should grow as $1/\cos\Phi$ when turning the field rotation plane away from the open FS sheets since only the field component parallel to the 1D sheets is responsible for the effect. The experimental data can only be fitted with this $1/\cos\Phi$ dependence if one assumes that the open sheets are running in a tilted direction compared to the one predicted from the band-structure calculations. This, however, is clear evidence for a reconstructed FS in the α-phase compounds with M = K, Rb, and Tl. Based on the tilt angle Φ_0, a nesting vector Q was proposed which should modify the original FS in the drastic way as shown in Fig. 4.15 [313]. In the FS of the metallic state one of the possible nesting vectors for the M = Rb and M = Tl salts, $Q = \frac{1}{6}k_a + \frac{1}{3}k_c + \frac{1}{6}k_b$, is plotted [312, 313]. For the M = K salt a nesting vector of $Q = \frac{1}{8}k_a + \frac{3}{8}k_c + \frac{3}{16}k_b$ was proposed [311, 312]. Other groups have in principle found the same result. The direction of the nesting vector, however, varies between 20° and 30° for α-(ET)$_2$KHg(SCN)$_4$ [290, 309].

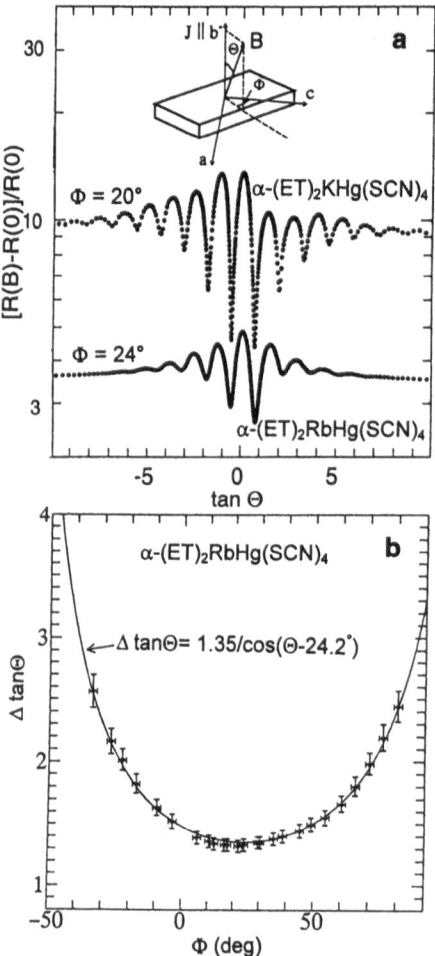

Fig. 4.14. (a) Angular dependent magnetoresistance oscillations (AMROs) in α−(ET)$_2M$Hg(SCN)$_4$ with M = K and Rb as a function of $\tan\Theta$ at angles Φ where maximal amplitudes were observed. The inset illustrates the experimental geometry. (b) Azimuthal dependence of the oscillation period $\Delta\tan\Theta$ in the M = Rb salt. From [312]

In the reconstructed FS the original quasi-1D bands disappear under the SDW potential but a new pair of open sheets tilted by Φ_0 against the original ones appear. Small lens-like closed orbits (λ_1) remain which may be separated by only small energy barriers from the open orbits. Therefore, it is suggested

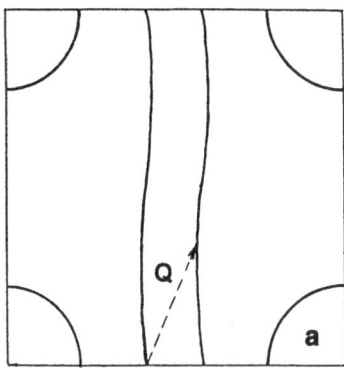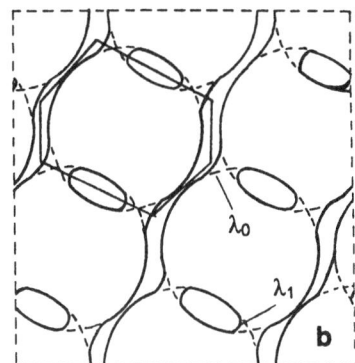

Fig. 4.15. FS of α-(ET)$_2M$Hg(SCN)$_4$ in (a) the metallic state with the proposed nesting vector Q and (b) the reconstructed FS in the low-temperature SDW state. From [313]

that the observed SdH and dHvA oscillations in the SDW state are due to MB orbits (λ_0) following the original quasi-2D closed orbits from which they emanate.

If the reconstructed FS is really due to the SDW transition which manifests itself in the resistivity hump at T_A and the kink field a drastic change of the AMRO signal should be observable above and below the phase transition. Indeed, an experiment in α-(ET)$_2$KHg(SCN)$_4$ at 1.5 K below ($B = 12$ T) and above ($B = 24$ T) the kink field proved the change in nature of the AMRO from 1D magnetoresistance dips to 2D maxima, respectively [310]. The same change in character of the AMRO was observed at 14 T above and below the magnetic ordering temperature at 8 K [311]. These results confirm the proposed FS reconstruction to be presumably due to a SDW transition.

It is somewhat surprising why some of the α-phase salts undergo a SDW phase transition while others remain stable in their high-temperature electronic structure. Since this effect depends on the overlap integrals of the electronic wave functions there might be a very subtle influence of the lattice parameters. Therefore, SdH investigations have been started under pressure. In α-(ET)$_2$NH$_4$Hg(SCN)$_4$ measurements under hydrostatic pressure have revealed an additional low oscillation frequency of approximately 40 T at ~ 3 kbar [314]. Since no slow oscillations have been seen in M = K, Rb, and Tl under the same conditions it is assumed that in M = NH$_4$ the oscillations result from pressure-induced nesting of the open orbits which are already nested at ambient pressure in the other salts. The nesting condition for M = NH$_4$, however, is incomplete since no SDW state appears under normal conditions. Therefore, the small orbits corresponding to the low frequency are attributed to the imperfect nesting of the 1D bands in this material [314].

In a further study of α-(ET)$_2$KHg(SCN)$_4$ under uniaxial stress perpendicular to the conducting planes a similar low frequency of ~ 52 T in the SdH effect has been reported [315]. This is attributed to the destruction of the perfect nesting in this compound and the concomitant appearance of a small orbit. In addition to this, the same group reported a stress-induced superconductivity in α-(ET)$_2$KHg(SCN)$_4$ [316]. Under the uniaxial pressure of 1.6 kbar the resistance of the sample at 0.5 K was strongly reduced; however not to zero. Under a magnetic field the resistance recovered quickly close to the value at $P = 0$. Concomitant with the appearance of the resitivity drop an increase of the effective cyclotron mass to $(1.7 \pm 0.1)\, m_e$ was observed. In α-(ET)$_2$NH$_4$Hg(SCN)$_4$ uniaxial stress perpendicular to the ET planes increased the superconducting transition from hardly above 1 K up to approximately 2.5 K at 3.8 kbar [316].

Table 4.1. Band-structure parameters of α-(ET)$_2X$ obtained from dHvA and SdH experiments. A_F is the FS area and A_{BZ} denotes the 2D Brillouin zone area which is approximately 40.7 nm^{-2} and only slightly changing for the different compounds; F_0 is the fundamental oscillation frequency at $\Theta = 0$; m_c/m_e is the cyclotron effective mass extracted from the temperature dependence of the oscillation amplitude; $g\mu_b$ is obtained from the angular dependence of the spin-splitting zeros

$X =$	F_0 (T)	A_F/A_{BZ}	m_c/m_e	$g\mu_b$	Ref.
NH$_4$Hg(SCN)$_4$	567	13.3	2.6[a]	4.45	[278]
TlHg(SeCN)$_4$	652	15.5	2.0	3.7	[287]
KHg(SeCN)$_4$	670	15.5	2.0	_[b]	[144]
KHg(SCN)$_4$	660	15.5	1.4[c]	3.07[d]	[300]
RbHg(SCN)$_4$	654	15.5	_[b]	_[b]	[307]
TlHg(SCN)$_4$	660	15.5	1.4	_[b]	[305]

[a]In [314] a value of 1.4 is reported. The stated value, however, was verified within error bars by [317].
[b]No data reported.
[c]In [290] a value of 1.9 is stated. Above the kink field a value of ~ 2.7 is reported [291].
[d]Above the kink field $g\mu_b \approx 4$ is found [300].

An interesting point when comparing the α-(ET)$_2X$ salts are the different effective masses and FS areas summarized in Table 4.1. The only superconducting compound with $X =$ NH$_4$Hg(SCN)$_4$ has the smallest FS area and the largest effective cyclotron mass. From the angular dependence of the spin-splitting zeros this salt further shows the largest value of $g\mu_b$ resulting in $\lambda = 0.17$ for $g = 2$. All other salts have within error bars the same FS area, approximately 16% larger than in the former. Consequently the effective cyclotron mass is smaller which would be expected in a free-electron picture where $\epsilon_F = \hbar^2 k_F^2/2m$. However, the mass reduction is 30% for $X =$ TlHg(SeCN)$_4$ and almost 50% for the other compounds in their SDW state. For the former salt this effect can be understood with the concomitant reduction of the electron–phonon interaction. As mentioned, the estimate of

λ from spin-splitting zeros yields $\lambda \approx 0.08$ which accounts for the observed mass difference within error bars.

In a more recent calculation [169] it was shown that the band structure of the α-phase salts strongly depends on the way the overlap integrals are taken into account (see Sect. 2.3.2). The two Fermi surfaces shown in Fig. 2.21 differ remarkably from each other. Therefore, it might also be plausible that the FS difference between the superconducting and non-superconducting salts is caused by slight changes in the electronic overlaps. Since in the SDW state of the relevant compounds presumably different parts of the FS are traversed by the charge carriers the reduced masses are not surprising. Indeed, above the kink field in the normal metallic state in the first experiments a mass enhancement up to $m_c \approx 2.0\,m_e$ was assumed [300]. Later work, however, revealed no change of m_c between both states. This discrepancy needs further investigations. Especially the angular dependence of the SdH or dHvA amplitude in the metallic state with possible spin-splitting zeros would give useful information. If the value $g\mu_b$ changes from the SDW to the metallic state this would imply a change of either m_c, λ or the electron–electron interaction. The latter mechanism was proposed as being responsible for the extremely small effective masses obtained from cyclotron resonance experiments in α-$(ET)_2KHg(SCN)_4$ [318]. In this work two masses of $m_{CR1} = 0.4\,m_e$ and $m_{CR2} = 0.94\,m_e$ were found. In the light of the results discussed above, however, it is not clear why two cyclotron resonance masses occur. Nevertheless, when comparing these values to the effective cyclotron masses m_c between 1.4 and $2.0\,m_e$ the mass enhancement due to electron–electron interaction would be enormous. This enhancement, however, might explain the difference between m_c and the effective masses estimated from band-structure calculations which give values around $1\,m_e$.

In conclusion, the FS of the α-phase salts has been extensively investigated. The FS areas are in qualitative agreement with band-structure calculations. However, the dHvA, SdH, and AMRO experiments revealed a clear difference in the band-structure parameters between the compounds that show a DW transition and the others. It was found that the strongest electron–phonon interaction is present in the superconductor α-$(ET)_2NH_4Hg(SCN)_4$. Some indications have been seen that in addition a strong electron–electron interaction exists in α-phase compounds. Three salts are unstable against a SDW transition which is supposed to suppress the superconducting transition. AMRO experiments were able to determine the reconstructed FS and the nesting vector. The details of the FS topology in the SDW state are still unclear and remain to be clarified. The α phase is unique among the 2D ET materials in the sense that almost perfectly nested 1D FS sheets are present in addition to the 2D band. Therefore, these salts have an intermediate position between 1D and 2D materials.

4.2.2 β Phase

β-$(ET)_2X$ compounds were one of the first 2D superconductors discovered. As mentioned, the superconducting transition temperature is related to the linear anion length with some exceptions where disorder reduces or suppresses T_c. Special interest was focused on the different superconducting states in β-$(ET)_2I_3$, the first ambient pressure superconductor. To obtain a better understanding of how the T_c dependence is related to the electronic structure comparative FS investigations between different β-phase salts were made. The calculated FS of the β phase shown in Fig. 2.17 is the simplest among all 2D materials and experiments have confirmed this single-band structure. The FS cross section of β-$(ET)_2IBr_2$ obtained from AMRO of the slightly warped cylinder has already been shown in Fig. 3.3 [172]. Although the principal topology of the FS seems to be clear some unusual and not yet fully understood effects are observed. In particular, the high harmonic content in the magnetic quantum oscillations and an unexpected large value of $g\mu_b$ are features which might be related to an extremely 2D nature of the FS.

Some of the first SdH experiments were reported for β-$(ET)_2I_3$ and β-$(ET)_2IBr_2$. However, these results were somewhat controversial. One group stated SdH frequencies of $F_0 \approx 1600\,T$ for $X = I_3$ and $F_0 \approx 1700\,T$ for $X = IBr_2$ [319]. In both salts a beating behavior of the SdH oscillations was observed. However, the oscillations in these experiments were very weak and did not behave as predicted by (3.6), i.e., they are not growing in amplitude with increasing field. Indeed, other experiments showed controversial results which were confirmed by later measurements.

Investigations done at the same time as those mentioned before revealed, for β-$(ET)_2I_3$ in the low-T_c state, only one very low SdH frequency of $F_{low} \approx 110\,T$ [320, 321]. No signs of fast oscillations nor a beating behavior were observed. This result is, however, not compatible with the predicted FS topology either. The frequency does not grow like $1/\cos\Theta$ and is only visible over a narrow angular region. A similar behavior was seen in β-$(ET)_2IBr_2$ where $F_{low} \approx 50\,T$ and an additional SdH oscillation of $F_0 \approx 3900\,T$ was reported [30, 321, 322]. The high frequency is consistent with the predicted area of band-structure calculations and will be discussed later in detail. The slow oscillation was ascribed to a "teapot"-like FS [322]. This, however, has not been confirmed by any calculation nor by dHvA experiments and remains to be clarified.

The origin of why in β-$(ET)_2I_3$ no high frequency corresponding to the expected FS area is observed seems to be the disorder within the CH_2 end groups and concomitant within the anion layer which occurs around $175\,K$ [323]. This disorder presumably causes too much scattering of the charge carriers with, therefore, a strongly enhanced Dingle temperature which hampers the observation of magnetic quantum oscillations. This assumption stems from the fact that SdH and dHvA oscillations are observable in the high-T_c state of β_H-$(ET)_2I_3$.

Figure 4.16 shows SdH results obtained for two different crystals which were cooled down under pressures of up to 1 kbar [324]. Fast SdH oscillations of 3730 T in agreement with the calculated FS area are visible in both samples with a huge and anharmonic amplitude as shown in the inset of Fig. 4.16a. This magnetoresistance shape is similar to the behavior found in α-(ET)$_2$TlHg(SeCN)$_4$ shown in Fig. 4.9. The large amplitude of the oscillations is attributed to a highly 2D FS. The degree of anharmonicity is sample dependent and, in addition, a beating behavior in the SdH signal is visible which indicates that the FS is corrugated. Hence, for the experimental accessible field range a number of Landau tubes are still cutting the FS simultaneously. The difference between the two extremal areas was found to

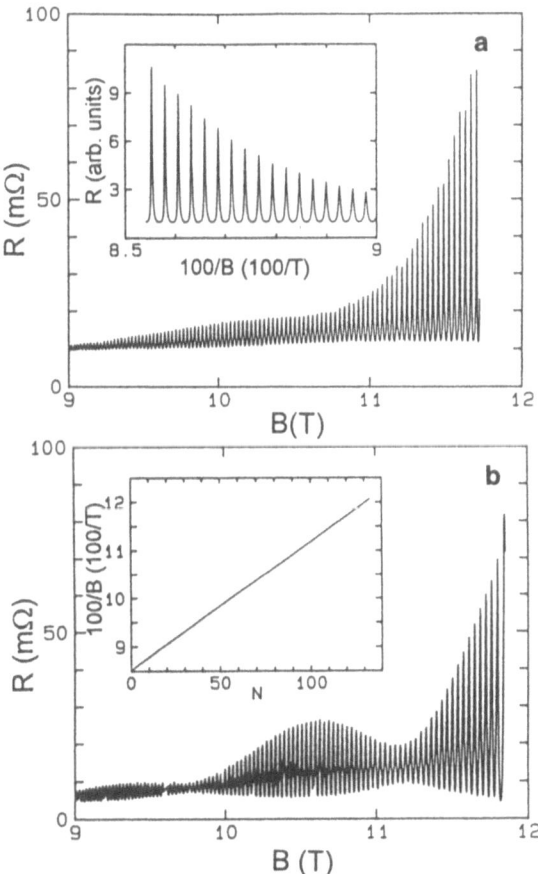

Fig. 4.16. Magnetoresistance of β-(ET)$_2$I$_3$ in the high-T_c state for two different samples (a) and (b). Inset (a) shows the strong anharmonicity. Inset (b) reveals the good periodicity of the oscillations in $1/B$. From [324]

be $\Delta F \approx 74\,\mathrm{T}$ for the only applied field direction, B perpendicular to the ET planes. In summary, the first and to date only SdH result in β_H-$(ET)_2I_3$ revealed a FS area consistent with band-structure predictions. The origin of the large anharmonicity of the oscillating magnetoresistance signal, however, is unclear. It might be an artifact of the resistivity data which may be reduced or even absent in the real SdH signal, namely in $\tilde{\sigma}/\sigma$ (see Sect. 3.2). In addition, the angular dependence of the magnetic quantum oscillations had not been investigated.

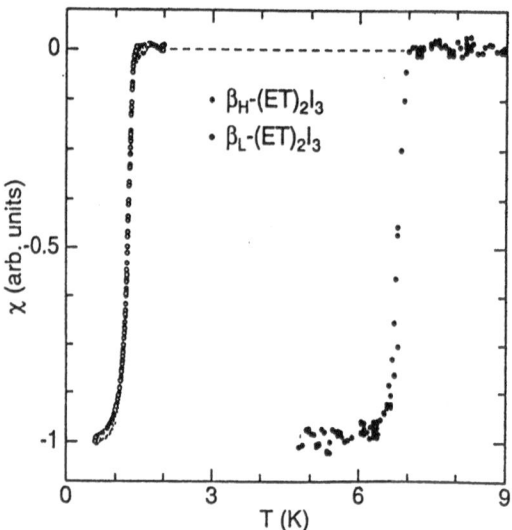

Fig. 4.17. Temperature dependence of the ac susceptibility of one β-$(ET)_2I_3$ crystal in the high-T_c state and after thermal cycling of the sample up to room temperature in the low-T_c state

In a very recent experiment the first observation of dHvA oscillations in β_H-$(ET)_2I_3$ under ambient pressure was successfully realized [325]. Because of the experimental difficulty in measuring dHvA oscillations under pressure a different approach to prepare the high-T_c phase was chosen. As mentioned already in Sect. 2.3.3, the β_H phase can be stabilized by moderate pressure of a few hundred bar when cooling down the samples through the transition below ~ 125 K. As long as the samples stay below this temperature the crystals remain in the ordered phase even if the pressure is released. Consequently, the dHvA sample was cooled down to liquid nitrogen temperature under $\sim 1\,\mathrm{kbar}$ He pressure, then the pressure was released and the sample mounted under liquid nitrogen into the dHvA probe which was then quickly transferred into a toploading cold ^{3}He cryostat. The T_c of the sample measured by ac susceptibility under ambient pressure was 7 K, proving the high-T_c phase of the crystal. Figure 4.17 shows the measurements of the sample in the β_H state and of the same crystal after warming up to room temperature. The T_c in the β_H state is somewhat smaller than the commonly stated 8 K which, however, has been obtained from the midpoint of resistivity curves with a relatively broad transition of ~ 2 K [216]. Only slightly above 7 K was the resistivity found to be zero. In another magnetic measurement, namely Meissner effect experiments under 1.5 kbar, $T_c \approx 6$ K was found at this pressure [135].

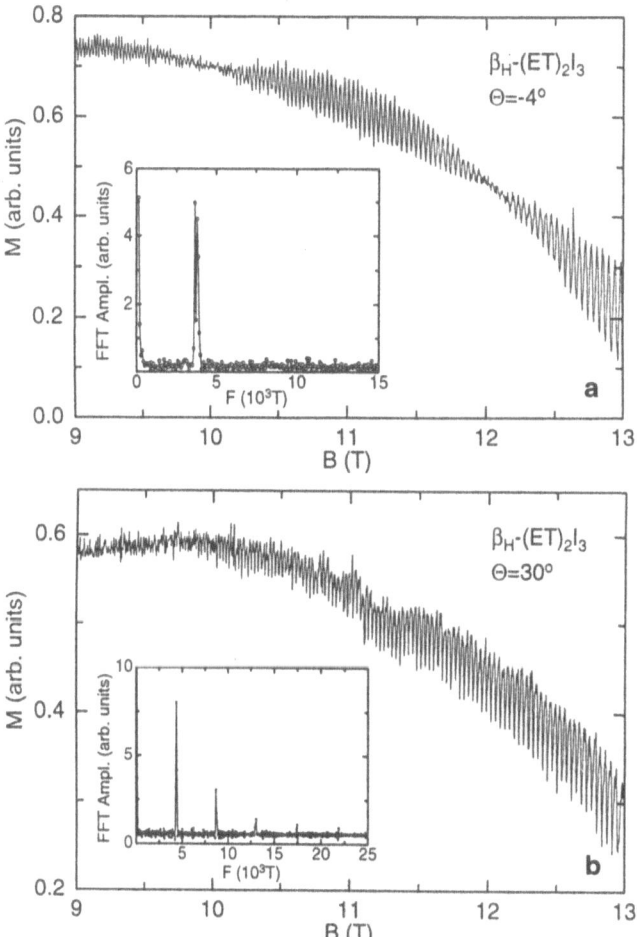

Fig. 4.18. dHvA signal for the β_H phase of $(ET)_2I_3$ for (a) $\Theta = -4°$ and (b) $\Theta = 30°$ at $T \approx 0.5$ K. The insets show the corresponding Fourier transforms with up to five harmonics for case (b)

Figure 4.18 shows the dHvA signal of the sample at $T \approx 0.5$ K for fields applied at $\Theta = -4°$ and $\Theta = 30°$. In reasonable agreement with the earlier SdH results, an average dHvA frequency of $F_{av} = (3805 \pm 10)$ T at $\Theta = -4°$ with two clearly visible nodes is found. The frequency difference is $\Delta F \approx 57\,\text{T}$, which is in rough agreement with the observations of the SdH measurements. However, in contrast to the SdH results at angles close to $\Theta = 0°$ the dHvA amplitudes are not unusually large and contain no visible higher harmonics as can be seen from the FFT of the data shown in the inset of Fig. 4.18a.

Systematic investigation of the angular dependence of the dHvA signal revealed the usual $1/\cos\Theta$ dependence but also a quite distinctive behavior of the dHvA amplitude, of $\Delta F(\Theta)$, and of the harmonic ratio. The $1/\cos\Theta$ angular dependence of F_{av} and of the effective cyclotron mass $m_{\text{c}}/m_{\text{e}}$ is shown in Fig. 4.19. Therefore, it can be concluded that the FS is almost two dimensional with only a small corrugation of the FS cylinder. To investigate this closer the frequency difference ΔF has been extracted from the experimentally observed nodes for different angles.

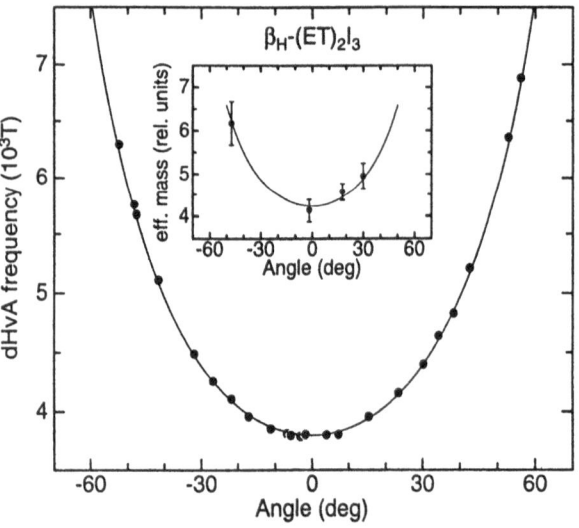

Fig. 4.19. Angular dependence of the average dHvA frequency and of the effective cyclotron mass (inset) of β_{H}-$(\text{ET})_2\text{I}_3$

Figure 4.20 shows the result obtained. The maximum frequency difference is found for $\Theta \approx 7°$ from where ΔF decreases in both directions. For $\Theta \approx 28°$ and $\Theta \approx -16°$ ΔF tends to zero. At these angles the oscillation amplitude is enhanced and the harmonic content of the signal is enormously increased. As an example Fig. 4.18b shows the result for $\Theta = 30°$ where in the FFT peaks up to the fifth harmonic were found. For some other measurements near 30° up to eight harmonics were visible. The behavior of $\Delta F(\Theta)$ is due to the warped FS (see Fig. 2.22) and can quantitatively be described with the model presented in Sect. 2.3.2. As in the derivation of (3.18), one has to take into account the triclinic nature of the crystal structure. This results

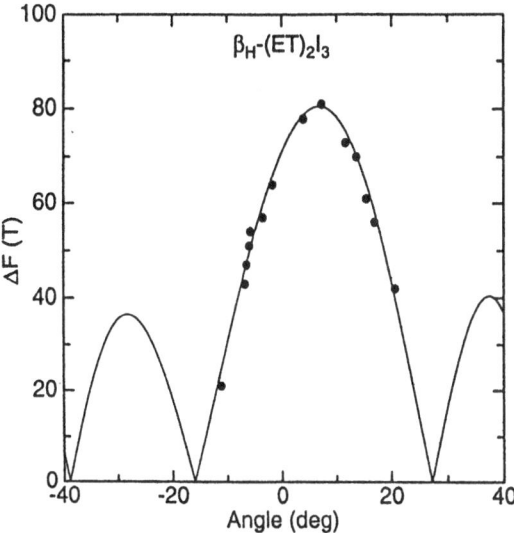

Fig. 4.20. Angular dependence of the dHvA frequency difference in β_{H}-(ET)$_2$I$_3$. The solid line is a fit according to (4.6)

in an interlayer transfer integral which has in-plane components or, in other words, the warped structure of the FS is tilted with respect to the normal of the conducting planes (see also Sect. 3.3). Depending on the azimuthal angle Φ of the field rotation plane, the angular change of ΔF becomes asymmetric and (2.11) has to be modified to

$$\Delta F \cos \Theta = \Delta F_{\mathrm{max}\Phi} J_0(c' k_{\mathrm{F}} \tan \Theta - u_{\|} k_{\|}), \qquad (4.6)$$

where for simplicity a circular cross section of the FS cylinder has been assumed. The zeros of this function for an arbitrary basal plane have already been given in (3.18). The prefactor for $J_0(x) = 1$ depends on the field rotation plane and is given by

$$\Delta F_{\mathrm{max}\Phi} = 4\, F_0 \frac{t}{\epsilon_{\mathrm{F}}} \cdot \frac{J_0(u_{\|} k_{\|})}{J_0(u_{\|} k_{\|})}. \qquad (4.7)$$

From Fig. 4.20 the asymmetry of ΔF is obvious. The value of $\Theta = 7°$ is somewhat smaller than the crystal angle between the c direction and the normal to the ab plane, which is tilted by 8.7° away from c as obtained from the room temperature lattice parameters (see Table 2.1). It can be assumed that the direction vector $h = (u_x, u_y, c')$ of the interlayer transfer integral coincides with the crystal lattice parameter c. Therefore, the rotation plane in the experiment was approximately 37° tilted away from $u_{\|}$. Because of the complicated handling of the sample no other field rotation plane was investigated. This was, however, done for the isostructural salt β-(ET)$_2$IBr$_2$ discussed later.

With $k_F = \sqrt{2eF_{av}/\hbar}$ and the room-temperature interlayer distance c' the angular dependence of ΔF can almost be described with (4.6). The solid line in Fig. 4.20 has been obtained with (4.6) assuming an elliptical FS area with an elongated k_F (15% compared to a circular FS) in the rotation plane. The only further fit parameter is $t/\epsilon_F \approx 1/175$. More complete information about the in-plane FS shape could fairly easily be obtained by AMRO measurements. In principle, this information is also extractable from dHvA or SdH experiments through the angles where ΔF is zero and using k_F as a fitting parameter.

At the angles where $\Delta F = 0$ the amplitude of the dHvA oscillations was enlarged. For these angles the FS cross section does not change along B. Therefore, all electrons of the metal contribute to the dHvA oscillations. This means that the curvature factor of the Lifshitz–Kosevich formula $|\partial^2 a_F/\partial\kappa^2|^{-1/2}$ becomes maximal. In addition to this expected behavior, however, hysteretic effects of the oscillating magnetization and an unusually high harmonic content of the dHvA signal could be observed which is visualized by the FFT of the data shown in the inset of Fig. 4.18b. The expected harmonic ratio for the maximum amplitudes M_r of the harmonic r to the harmonic $r + 1$ (that is for spin-splitting factors R_S equal to 1) can be extracted from the Lifshitz–Kosevich formula (3.6) to

$$\frac{M_r}{M_{r+1}} = \frac{\sqrt{r+1}}{\sqrt{r}\exp(-\alpha\mu_c T_D/B)}\frac{\sinh[(r+1)\alpha\mu_c T/B]}{\sinh(r\alpha\mu_c T/B)}. \tag{4.8}$$

With an estimated[4] $T_D = 0.4\,\mathrm{K}$ and $m_c = (4.1 \pm 0.1)\,m_e$ the harmonic ratio for M_1/M_2 is ~ 130. Therefore, the maximal amplitude of the second harmonic would be smaller by this factor compared to the maximal fundamental amplitude. This, however, would mean that M_2 and of course also all higher harmonics are always below the noise level for the present crystal and experimental sensitivity. The fact that harmonics up to the eighth order are visible at the angular range where $\Delta F = 0$ shows the invalidity of (3.6) for these angles where only one FS area like in a perfect 2D material exists.

For a perfectly 2D material the usual assumption of a constant field independent chemical potential has to be revised. If in an increasing field the FS is just between two successive Landau levels the chemical potential has to shift with the Landau cylinders and finally jump back for increasing field to the next lower level.[5] An analytical formula for the dHvA effect of a 2D electron gas taking into account finite temperature and fields is given in [326]. The applicability of this formula could not be tested quantitatively. Qualitatively, however, the predicted increased harmonic content and a drastically changed shape of the oscillations was observed.

---

[4] Because of the nodes only for $\Delta F = 0$ the Dingle temperature could be determined from the dHvA amplitude dependence over an larger field range. There, however, the Lifshitz–Kosevich formula might no longer work.

[5] In the limit where only a few Landau cylinders are inside the FS the well-known quantum Hall effect [74] should be visible.

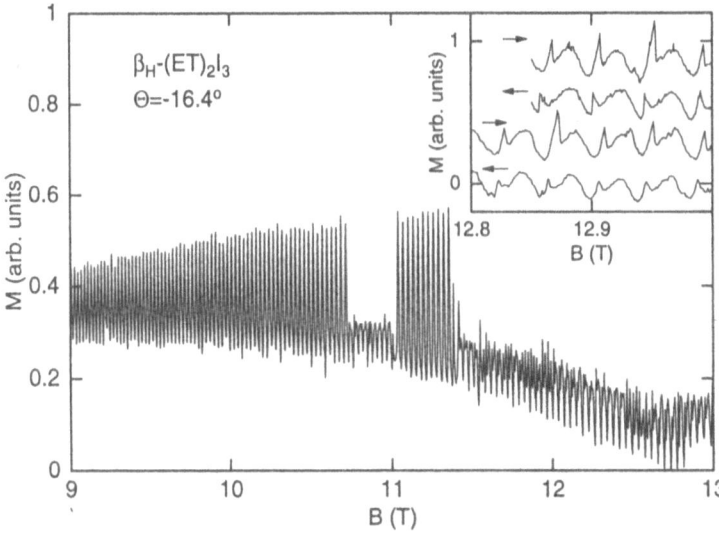

Fig. 4.21. dHvA signal of β_H-$(ET)_2I_3$ at $\Theta = -16.4°$ where $\Delta F \approx 0$. During the field sweep sudden strong reductions of the dHvA signal appeared. The inset shows the hysteresis of the reduced dHvA signal for two up and two down sweeps (indicated by the arrows). For clarity the signals are shifted vertically with respect to each other

In addition to the strong anharmonic content of the dHvA signal at the angular range where $\Delta F = 0$ a sudden collapse of the oscillation amplitude and hysteretic behavior could be observed. Figure 4.21 shows as an example the dHvA signal at $\Theta = -16.4°$. During the up-sweep at certain field ranges the strongly anharmonic oscillations reduce by approximately a factor of three in amplitude without a change in the dHvA frequency. The inset shows the hysteretic behavior of the signal at the field range where the amplitude is reduced. A possible explanation for this effect is magnetic interaction (see also the discussion for β''-$(ET)_2AuBr_2$ below) and the formation of a domain structure. For the nearly perfect 2D extremal shape of the FS at these special angles the strong harmonic content of the dHvA signal may result in a change of magnetization with field, dM/dH, which is larger than 1. For plate-like samples like the one investigated the formation of so-called Condon domains [327] is expected. These domains depend on field and temperature. Indeed, in some experimental runs the oscillation amplitude could be enlarged by briefly heating the sample by a few hundred mK [325]. Although the features observed in β_H-$(ET)_2I_3$ at the angles where $\Delta F = 0$ seem to be consistent with the formation of domains a definite proof is lacking.

Due to the large effective mass of $4.1\,m_e$ and the measured $1/\cos\Theta$ dependence a large number of spin-splitting zeros vs angle are anticipated and, indeed, observed (see Sect. 4.2.1). In the dHvA experiment three definite angles with almost vanishing fundamental dHvA amplitudes were discovered.

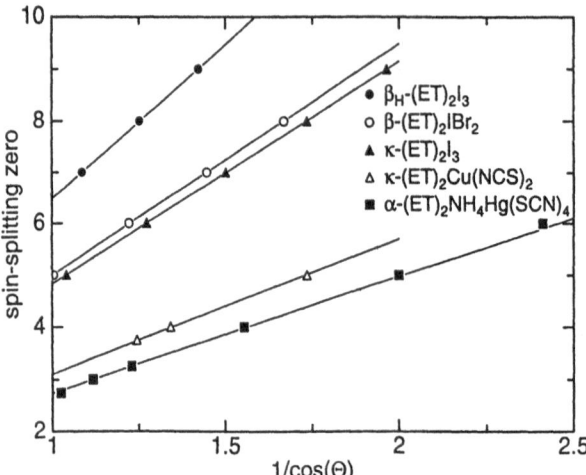

Fig. 4.22. Observed spin-splitting zeros of the fundamental and the second harmonic dHvA amplitude for five different organic superconductors vs $1/\cos\theta$. The solid lines are linear fits described in the text

With the assumption of a $1/\cos\Theta$ dependence of μ_b the angles Θ_n are given by

$$n = \frac{g\mu_{b0}}{2} \cdot \frac{1}{\cos\Theta_n} + \frac{1}{2}. \tag{4.9}$$

Figure 4.22 shows a plot of n vs $1/\cos\Theta$ for the observed spin-splitting zeros in five different organic superconductors. The slopes, thereby, directly give $g\mu_{b0}/2$. The intercepts of the linear fits are forced to go through $n = 1/2$ determining the absolute value of n. For β_H-$(ET)_2I_3$ $g\mu_{b0} = 11.96$ is obtained. With the ESR value of $g \approx 2$ this would give a μ_{b0} larger than μ_{c0} which is impossible. This either means that g is much larger than 2 or the determination of μ_c is already influenced by the invalidity of the 3D description of Lifshitz and Kosevich for the present quasi 2D material. A similar effect has been observed for at least two other materials, β-$(ET)_2IBr_2$ and κ-$(ET)_2I_3$, also shown in Fig. 4.22. This will be discussed later in more detail.

In order to verify that the disorder in the anion layer and the terminal ethylene groups enhances the electron scattering and prevents the observation of dHvA oscillations the sample was warmed up to room temperature and afterwards cooled down and measured again. T_c was reduced to $\sim 1.35\,\text{K}$ as shown in Fig. 4.17 and no dHvA oscillations were longer visible. Since the overlap integrals and the FS should not be influenced by the disorder [154] the most plausible explanation for this observation is a drastic increase of the Dingle temperature in the sample. The fact, however, that this disorder also affects the superconducting transition temperature is unusual. Assuming a phononic superconductivity mechanism, one can only speculate that the disorder changes some phonon modes which might be important for the superconductivity.

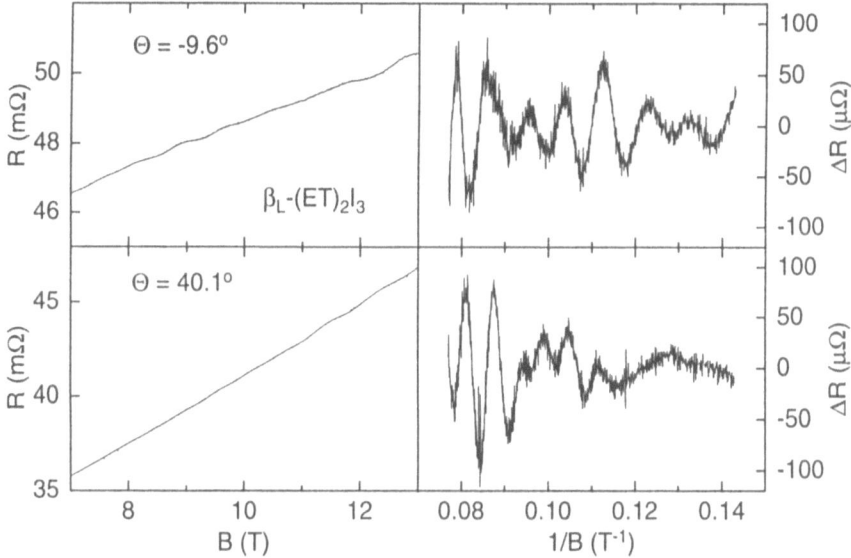

Fig. 4.23. The left panels show the field dependence of the resistivity of the low-T_c phase of β-(ET)$_2$I$_3$ for two different angles. The right panels show the oscillatory contributions of the magnetoresistance after subtraction of smooth backgrounds at the same angles vs reciprocal field

Interestingly, however, measurements of magnetoresistance of another β-(ET)$_2$I$_3$ sample under ambient pressure revealed weak SdH oscillations.[6] Figure 4.23 shows the magnetoresistance between 7 and 13 T for two angles. On an almost linearly increasing background resistance weak oscillations are visible. After subtraction of this background signal the SdH oscillations can be seen more clearly in the right panels of Fig. 4.23. The SdH frequency is approximately 110 T in good agreement with the value stated in [320, 321]. Clear nodes in the oscillating signal are visible, suggesting the existence of more than one extremal FS area. For small angles (Θ less than $\sim 35°$) at the nodes no phase shift of the oscillations is observed, whereas for higher angles this shift is present as expected for only two slightly different SdH frequencies. This suggests a complicated FS topology or possibly interference effects in the low-T_c state. The average SdH frequency is independent of angle up to $\sim 40°$ where a suddenly increased frequency of ~ 185 T is found. The effective mass for the 110 T frequency is approximately $1\,m_e$. The determination of the Dingle temperature is difficult due to the nodes but is estimated to be of the order of 1 K.

The above results are in contrast to the predictions of band-structure calculations and to the FS topology found in the high-T_c state. As mentioned,

---

[6] The oscillation frequency is presumably too slow to be detected by the modulation-field method applied in the dHvA experiments mentioned.

a changed FS in the disordered phase is unlikely. Possibly the low-frequency oscillations are present in the high-T_c state as well but could be masked by the extraordinary large amplitude of the high-frequency oscillations in the SdH experiment [324] or by the reduced sensitivity for low-frequency oscillations in the modulation-field dHvA experiment [325]. Further studies are necessary to elucidate this mysterious picture.

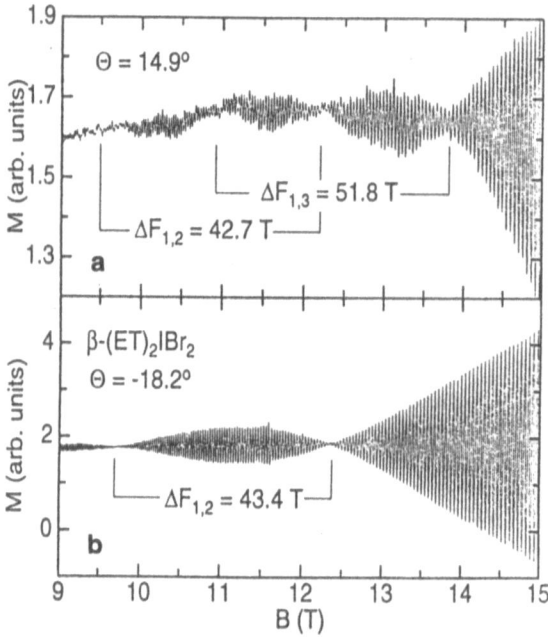

Fig. 4.24. dHvA signal for β-(ET)$_2$IBr$_2$ at $T \approx 0.5$ K for **(a)** $\Theta = 14.9°$ and **(b)** $\Theta = -18.2°$

The band-structure parameters of β-(ET)$_2$IBr$_2$ extracted from SdH and dHvA experiments are similar to those just described for the high-T_c state of β_H-(ET)$_2$I$_3$. Figure 4.24 shows the dHvA signal measured at $T \approx 0.5$ K for two different angles [328]. For both angles a clear beating behavior can be seen. In the measured field range between 9 and 15 T for $\Theta = 14.9°$ four nodes are visible, two of them with nearly zero remaining oscillation amplitude at ~ 9.5 T and ~ 12.2 T, and two with finite amplitude at ~ 10.9 T and ~ 13.8 T.[7] This and the relative positions of the nodes can only be understood with the existence of four extremal areas leading to four slightly different frequencies, F_1 to F_4. The differences are $\Delta F_{1,2} = F_1 - F_2 = \Delta F_{3,4} \approx 42.7$ T and $\Delta F_{1,3} = \Delta F_{2,4} \approx 51.8$ T. For $\Theta = -18.2°$ only two nodes remain, yielding a frequency

---

[7] Reanalysis of the data has yielded slightly changed the difference frequencies compared to the published values in [328].

difference of $\Delta F \approx 43.4\,\text{T}$. The amplitude of the dHvA signal is a factor of 10 larger than for $\Theta = 14.9°$.

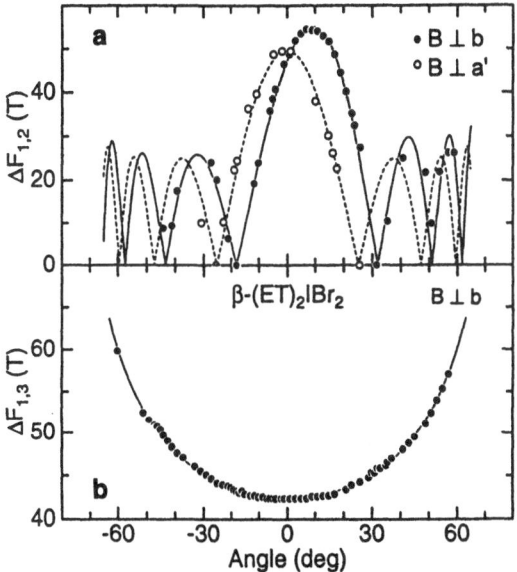

Fig. 4.25. Angular dependence of the beating frequencies (a) $\Delta F_{1,3}$ and (b) $\Delta F_{1,2}$ in β-(ET)$_2$IBr$_2$. The lines are fit curves described in the text

Systematic investigation of the angular dependence of the dHvA oscillations revealed that the frequency difference $\Delta F_{1,3}$ depends on the azimuthal angle of the field rotation plane and can be described perfectly by the warped cylinder model (4.6) with the only fit parameter $t/\epsilon_F \approx 1/280$ as shown in Fig. 4.25 [328]. This is the same behavior observed and described for β_H-(ET)$_2$I$_3$. Figure 4.25a shows the experimental points of $\Delta F_{1,3}$ for two field rotation planes. For $B \perp b$ the maximum asymmetry is found. Consequently, for $B \perp a'$, i.e., perpendicular to b, the frequency difference becomes symmetric around $\Theta = 0°$. The solid lines are the fits according to (4.6). This behavior, therefore, is quite well understood and has to some extent already been observed in SdH measurements [30, 322]. The direction and the angle of the highest asymmetry gives the in-plane component $u_{\parallel}$ of the transfer integral. It coincides well with the triclinic angle $\sim 8°$ and with the value obtained from the AMRO measurement described in Sect. 3.3 [172]. As distinct from dHvA measurements, AMRO experiments can only detect the presence or absence of a corrugation. The amplitude of warping, however, cannot be extracted. This is only possible when nodes in dHvA or SdH oscillations are found. Nevertheless, AMRO experiments can map out the in-plane FS topology and the direction of $u_\parallel$ in a much easier way than SdH or dHvA measurements could do this (see Fig. 3.3).

The remaining and still unresolved question is the origin of the frequency difference $\Delta F_{1,2}$ in β-$(ET)_2IBr_2$. The smooth angular dependence shown in Fig. 4.25b can be described by $\Delta F_{1,2} = (24.6 + 17.75/\cos\Theta)$ T. These nodes were only observed in dHvA experiments on different samples of the Argonne group but seem not to be present in Russian samples [329]. A possible twinning in the former seems to be highly unlikely since then $\Delta F_{1,2}$ would grow much faster with angle than observed and would depend on the field rotation plane. In addition, the nodes are always almost exactly zero which would mean that exactly equally sized parts of the sample with the same effective masses are showing dHvA signals. Another possibility to explain the observed behavior are the magnetic interaction effects due to weak antiferromagnetism.

The non-zero amplitude of the $\Delta F_{1,3}$ nodes is due to a small difference in the effective masses for the small and the large orbit. In first approximation m_b is proportional to the FS area of the cylinder. Indeed, with the parameters for the sample shown, $m_{c0} = 4.0\,m_e$ (see Fig. 4.6), $T = 0.5$ K, $T_D = 0.5$ K, and with an effective mass difference of approximately 1.3% one can quantitatively reproduce the remaining amplitude of the nodes.

The angular change of the average frequency F_{av} is the one expected for a cylindrical FS with $F_{av} = (3842 \pm 10)$T$/\cos\Theta$ [328]. This value corresponds to approximately 53% of the first Brillouin zone and is in excellent agreement with the SdH value [30, 322] and the area obtained by AMRO experiments shown in Fig. 3.3 [172]. The discrepancy in the early work mentioned [319] where a FS area of approximately half this value was reported is unclear.

Due to the warped nature of the FS and the $1/\cos\Theta$ dependence of $g\mu_b$ the angular dependence of the dHvA amplitude is nonmonotonic as shown in Fig. 4.26. The points scatter appreciably because just the FFT maximum amplitude of sweeps between 12 and 15 T were taken regardless whether nodes were present in this field range or not. Extremely large dHvA amplitudes are obtained at angles where the Bessel function in (4.6) has zeros. At these points the curvature factor $|\partial^2 a/\partial\kappa^2|^{-1/2}$ becomes maximal. In addition, from the angles with vanishing dHvA amplitude the spin-splitting zeros shown in Fig. 4.22 with $g\mu_{b0} = 8.99$ are extracted. The solid line in Fig. 4.26 is obtained with this value and the curvature factor calculated from (4.6). The principal behavior of the absolute amplitude with the spin-splitting zeros and the extraordinary large amplitudes at $\Theta = 18°$ (see Fig. 4.24b) and at $\Theta = 31°$ are well reproduced.

As already discussed for β_H-$(ET)_2I_3$, the question remains why with the ESR g-value of 2.007 [330] a bare electron mass μ_b larger than μ_c would be extracted. If no mass-enhancement due to electron–phonon interaction is considered, i.e., $\lambda = 0$, a lower limit of the g value of ~ 2.25 would be obtained from the dHvA data for β-$(ET)_2IBr_2$. This might hint at antiferromagnetic fluctuations or an appreciable electron–electron interaction which

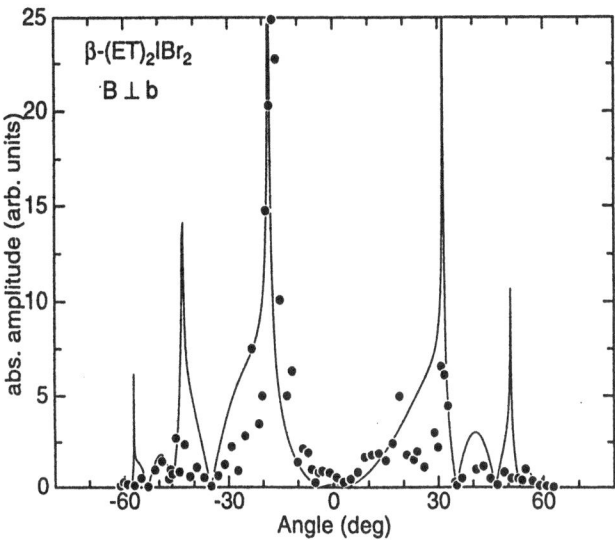

Fig. 4.26. Angular dependence of the amplitude of the dHvA oscillations in β-$(ET)_2IBr_2$. The solid line is a fit curve described in the text

was also assumed from the cyclotron resonance experiments in some other ET materials already discussed [282]. Since, however, a mass-enhancement due to electron–phonon interaction should be present, too, g could be even much larger. On the other hand, in the α-$(ET)_2NH_4Hg(SCN)_4$ discussed as in κ-$(ET)_2Cu(NCS)_2$ (see below), the value $g = 2$ is consistent with $\mu_b < \mu_c$ and results in a reasonable λ. Of course, the seemingly enhanced g value in the β-phase salts as well as in κ-$(ET)_2I_3$ (see next section) could indicate a similar enhancement for the above-mentioned materials as well. This, however, would mean that the extracted λ is only a lower limit. T_c then should be estimated with the McMillan formula (2.25). Another possibility for the different results could be the invalidity of the Lifshitz–Kosevich formula employed for the extraction of μ_c from the temperature dependence of the dHvA amplitude. The suggestion stems from the fact that in all the compounds where large values of $g\mu_b$ were observed enlarged harmonic contents in the dHvA signal were also found. In these materials, due to the simple one-band Fermi surface, all the electrons at the Fermi energy contribute to the SdH or dHvA signal. In contrast, for the α-$(ET)_2XHg(YCN)_4$ salts and κ-$(ET)_2Cu(NCS)_2$ additional open bands are present at the Fermi level (see Sect. 2.3.2). Therefore, in these latter compounds the chemical potential can remain constant with changing field since the 1D bands are acting like electron reservoirs which can provide carriers. For the more 2D compounds with only one FS the chemical potential jumps each time a Landau level crosses the Fermi energy and consequently the dHvA signal contains more higher harmonics than predicted by the 3D Lifshitz–Kosevich formula. How this influences the effective mass determination is not exactly clear but there might be a significant difference between the effective cyclotron mass determined via (3.7) and the real effective mass.

For one further β-phase salt, namely β-(ET)$_2$AuI$_2$, one early experiment claims the observation of both SdH and dHvA oscillations [331]. The results, however, are contradictory and remarkably different from those obtained for the other β-phase compounds and from the predicted FS of Fig. 2.17. The SdH data showed oscillations with a frequency of $\sim 20\,\mathrm{T}$ and an effective mass of $m_c = (2.0 \pm 0.5)\,m_e$. The dHvA frequency reported was $F \approx 290\,\mathrm{T}$ with $m_c = (0.3 \pm 0.1)\,m_e$ and showed no angular dependence of F up to $\pm 40°$. This would mean that β-(ET)$_2$AuI$_2$ has a spherical FS totally different from the other β-phase salts just discussed. Since this was the only experiment which reported the observation of magnetic quantum oscillations for this material a verification of the data would be desirable to clarify the discrepancies of the FS topologies.

Another material with molecular packing very similar to the β-phase structure is β''-(ET)$_2$AuBr$_2$. This compound was synthesized 1986 [332] and remains metallic down to at least $\sim 40\,\mathrm{mK}$. ESR measurements revealed a decrease of the spin susceptibility and a change of the g factor around $20\,\mathrm{K}$ [333]. In addition, the resistivity showed hysteretic behavior at this temperature region [333].[8] This suggests that an antiferromagnetic transition takes place at $\sim 20\,\mathrm{K}$.

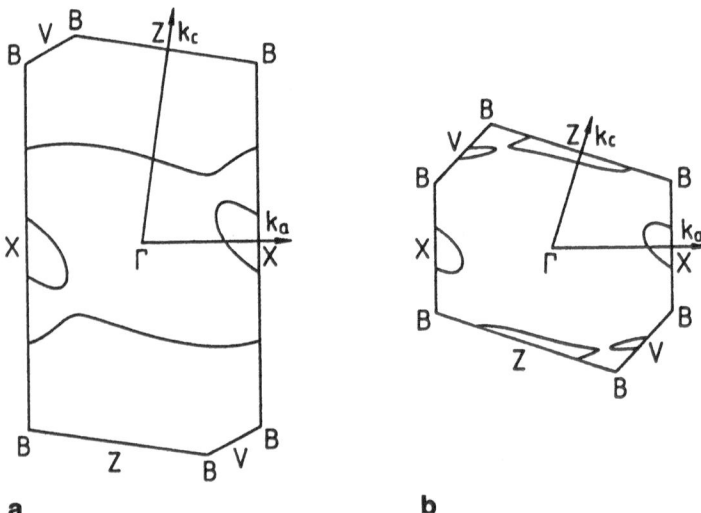

a   **b**

Fig. 4.27. (a) Calculated FS of β''-(ET)$_2$AuBr$_2$. From [335]. (b) Proposed reconstructed FS after the formation of a SDW. From [336, 337]

---

[8] In contrast to this observation, later results [334] did not show any signs of hysteresis in the temperature-dependent resistivity.

The main difference between the β'' structure compared to the β phase is the direction of the strong intermolecular interactions. Due to the smaller anion size the interaction directions are at $0°$, $30°$, and $60°$, respectively, instead of face-to-face ($90°$) overlaps [335]. The more complicated interstack interaction results in a more anisotropic band structure with 1D and 2D energy bands. There exists considerable disagreement between different band-structure calculations which might be caused by small differences in the transfer integral values [332, 335, 336]. One calculated FS based on the room temperature lattice parameters is shown in Fig. 4.27a [335]. Small 2D pockets occur around X and two 1D open sheets run perpendicular to the a direction. In contrast, the calculation of [332] (not shown) revealed a rather large closed orbit around the Γ point.

Both observed SdH and dHvA oscillations revealed a quite complex behavior and partially controversial results were reported. The SdH signal obtained at $T = 0.044\,\mathrm{K}$ for the field perpendicular to the conducting planes is shown in Fig. 4.28 [334]. The FFTs of the data done for different field regions reveal a large number of peaks. At lowest fields the two strong peaks at $41.5\,\mathrm{T}$, labeled α, and $139\,\mathrm{T}$, labeled β, are believed to belong to real cyclotron orbits. All other FFT peaks are interpreted as harmonics or combination frequencies. As can be seen from the FFT for the different field regions, the amplitudes of the main and combination peaks show a rather strange field dependence. This is interpreted as an indication for magnetic interaction as first introduced by Shoenberg [249] in the course of dHvA measurements on noble metals.

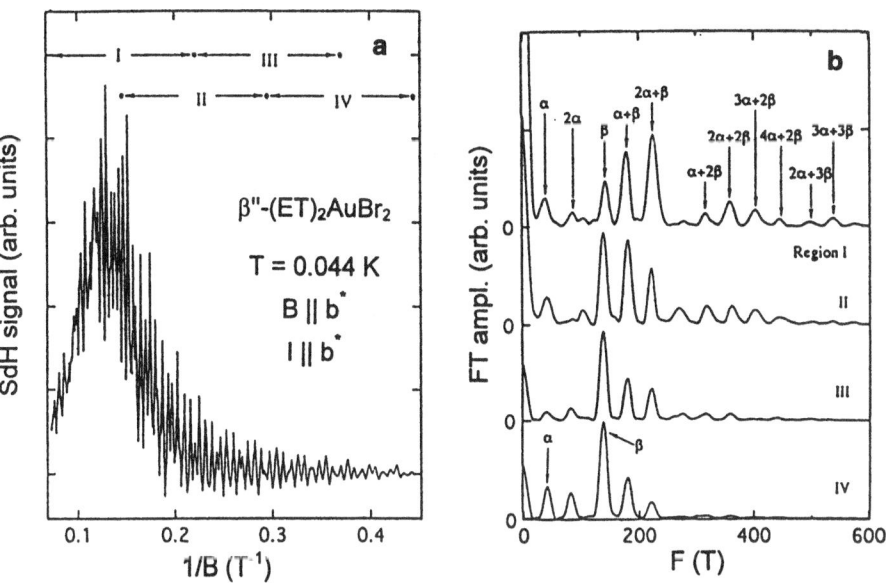

Fig. 4.28. (a) SdH effect in β''-$(ET)_2AuBr_2$ at $T = 0.044\,\mathrm{K}$ with (b) Fourier transformations for four different field ranges denoted by I to IV. From [334]

Magnetic interaction can be observed when the magnetization change with field, dM/dH, is comparable to 1, where H is the applied external field. At this point we have to distinguish between the fields H and B. In the Lifshitz–Kosevich formula one has to take into account that $M(B)$ is an implicit function determined by the relation $B = \mu_0(H + M)$. This magnetic interaction increases the harmonic content of the dHvA signal and further generates combination frequencies if more than one single frequency is present [249]. These combination frequencies become more pronounced for higher fields. Therefore, the relatively strong peaks of the Fourier transformation (α and β) at the lowest field region analyzed were claimed to correspond to real FS areas. They occupy $\sim 0.5\%$ and $\sim 1.7\%$ of the first Brillouin zone. This result is in disagreement with all band-structure calculations. One possible explanation for this discrepancy is the formation of a SDW at $\sim 20\,K$ where the ESR results suggested an antiferromagnetic phase transition [333]. Therefore, it was assumed that a $2c$ SDW occurs at this temperature which reconstructs the FS in a way shown in Fig. 4.27b [336, 337]. Due to imperfect nesting of the 1D FS sheets two new closed hole-like orbits are predicted. The sum of these orbits are supposed to be exactly equal to the original 2D electron-like pockets around X. Therefore, the peak in the FFT at $\alpha + \beta = 181\,T$ (see Fig. 4.28b) is proposed to correspond also to a real orbit. The experimentally obtained areas of these orbits, however, are only about 50% the predicted values [334].

When comparing the results of different groups a considerable spread of oscillation frequencies and effective masses can be found. In first SdH studies of one group frequencies around $140\,T$, $180\,T$, and $220\,T$ are reported [338]. Later the low frequency at $\sim 40\,T$ was also found [336, 337, 339]. This group originally predicted the described reconstruction of the FS. In this work, however, not the $140\,T$ but the $180\,T$ frequency was ascribed to the fundamental β orbit. Consequently, the $220\,T$ SdH frequency was assumed to correspond to the electron-like orbit. In a SdH and dHvA study frequencies around $50\,T$, $168\,T$, $218\,T$, and $268\,T$ were reported [340]. All these values are approximately a factor 1.2 larger than those of other groups. In addition, in this work no change of the frequency for one other measured angle at $\sim 25°$ was found. This is in sharp contrast to the perfect $1/\cos\Theta$ dependence reported otherwise [334, 336]. A possible explanation would be a strong misalignment of the sample in the dHvA study. The authors, however, claim that the angle might be off by $5°$ at the most [340].

A large discrepancy is reported for the values of the effective cyclotron masses of the different orbits. Due to the non Lifshitz–Kosevich behavior of the temperature-dependent dHvA amplitude in [336] only data above $\sim 1\,K$ were analyzed resulting in effective masses between $2\,m_e$ and $3.5\,m_e$. In [334] in the low-field region all data down to the lowest temperature followed the usual behavior and yielded values of $0.66\,m_e$, $0.92\,m_e$, and $1.2\,m_e$ for the α,

β, and $\alpha + \beta$ orbit, respectively. Within error bars similar values are given in [340] for the corresponding orbits.

None of the studies gives clear statements of observed spin-splitting zeros. In [336] a contour plot of the FFT frequency spectrum vs angle is given which showed the vanishing of a fundamental frequency only for very few angles. Therefore, the lower values of the effective masses seem to be more reasonable since otherwise more spin-splitting zeros should have been observed. Of course, the magnetic interaction complicates the situation and an analysis with, e. g., the harmonic ratio method is hardly possible.

In conclusion, the reconstructed FS shown in Fig. 4.27b seems to represent the measured data best. Similar to α-phase salts already discussed a SDW transition might change the high-temperature FS drastically. Some experimental discrepancies of the FS extremal areas and the effective cyclotron mass still exist which remain to be resolved.

4.2.3 κ Phase

The molecular ET packing arrangement denoted by the κ phase reveals organic superconductors with relatively high T_cs. A whole set of substances with T_c around 10 K including κ-$(ET)_2Cu[N(CN)_2]Cl$ with $T_c \approx 13\,K$ at $\sim 0.3\,kbar$ record to date [16] have been discovered. The considerable increase of T_c to a more easily accessible temperature range caused an enormous number of different investigations. The superconducting properties of the κ phase have been discussed in detail in Sect. 2.3.3.

κ-$(ET)_2Cu(NCS)_2$ has become one of the best studied 2D organic superconductors. This material was also one of the first organic metals where magnetic quantum oscillations were discovered [29]. Figure 4.29 shows the SdH oscillations of this first magnetoresistance measurement. Below $\sim 1\,K$ and above $\sim 9\,T$ clear oscillations are visible. In the following many other groups have reported SdH and dHvA oscillations in κ-$(ET)_2Cu(NCS)_2$. Increased magnetic field and, especially, the more systematic investigation of the angular dependence of the oscillations contributed to a nearly complete knowledge of the band structure for this material.

It is, however, rather surprising that κ-$(ET)_2Cu(NCS)_2$ was for a few years the only κ salt where magnetic quantum oscillations could be observed. Especially, in the high-T_c salts κ-$(ET)_2X$ with $X = Cu[N(CN)_2]Br$ and $X = Cu[N(CN)_2]Cl$ no oscillations even up to the highest accessible fields have been found under ambient pressure. Recently SdH oscillations under applied pressure of a few kbar have been observed in both materials (see below) [341]. The band structure of these salts, on the other hand, is expected to be almost exactly the same as that for κ-$(ET)_2I_3$ shown in Fig. 2.19. In this latter compound, however, very large SdH and dHvA oscillations in agreement with the predicted FS have been discovered. These results, as well as the unusual temperature and field dependence of the oscillations, will be discussed later in this section.

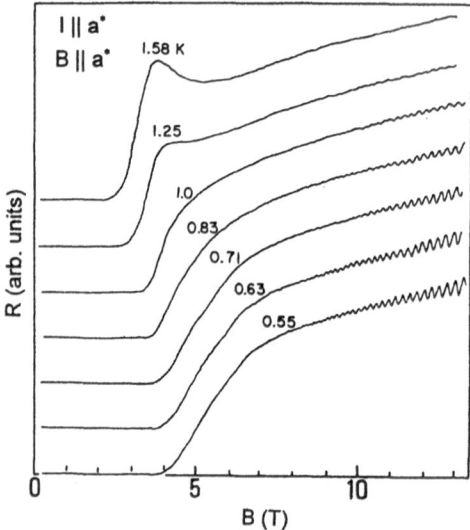

Fig. 4.29. Magnetoresistance of κ-$(ET)_2Cu(NCS)_2$ for different temperatures. At low temperatures and high enough fields clear SdH oscillations are visible. From [29]

Other attempts to extract the FS of κ-$(ET)_2Cu[N(CN)_2]Br$ were made by angular resolved photoemission [9] and by positron annihilation studies [342]. In the former experiment, however, the angular resolution was only 1/3 of the $\varGamma$–Z and 1/5 of the $\varGamma$–Y distance (see Fig. 2.19) due to the large unit cell. Nevertheless, the energy dependence revealed a broad onset of the photoemission intensity at the Fermi energy. The absence of a sharp Fermi edge and a corresponding very low density of states at the Fermi energy was also realized for κ-$(ET)_2Cu(NCS)_2$ [9, 10] and β-$(ET)_2I_3$ [9]. Whether this behavior is an indication of a possible deviation from the usual Fermi-liquid theory towards a Luttinger-liquid behavior [7] in a 2D system is unclear. The positron annihilation measurement [342] showed substantial discrepancies between the experimentally extracted FS and the predicted one. This might be due to a different FS in κ-$(ET)_2Cu[N(CN)_2]Br$ compared to the other κ-phase salts or due to the neglect of positron wavefunction effects in the experimental interpretations. Further studies are needed to clarify the real band structure of this material.

In the first SdH study of κ-$(ET)_2Cu(NCS)_2$ the $1/\cos\varTheta$ dependence of the SdH frequency with a period of $0.0015(1)\,T^{-1}$ ($F = (667\pm40)\,T$) has been reported [29, 343]. Subsequent investigations revealed a somewhat inconclusive picture on the exact value of the FS area within the 2D Brillouin zone [344, 345, 346, 347, 348]. In a more systematic dHvA study a high-quality sample was carefully oriented by a four-axis diffractometer. At $T = 0.5\,K$ the dHvA signal shown in Fig. 4.30 was found [278, 279]. For the shown field orientation perpendicular to the planes ($\varTheta = 0°$) above $\sim 6\,T$ clear dHvA oscillations with one frequency of $F = (598 \pm 3)\,\mathrm{T}$ are observed. The FFT of the data shows the spectral purity of the signal. dHvA oscillations were measured up to $\pm 65°$. The frequency followed the $1/\cos\Theta$ dependence exactly yielding a minimum frequency of $(598.5 \pm 1)\,\mathrm{T}$ [278, 279]. Later work verified this result quantitatively [32, 350, 351].

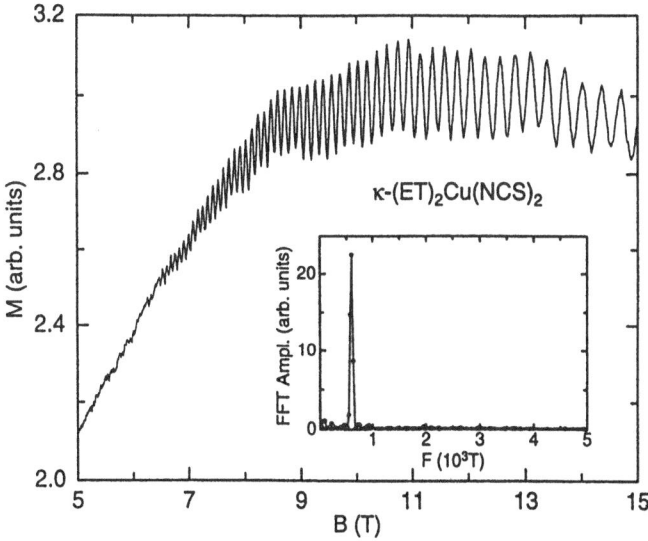

Fig. 4.30. Magnetization of κ-$(ET)_2Cu(NCS)_2$ at $T = 0.5\,\mathrm{K}$ for fields applied perpendicular to the ET planes. The inset shows the FFT of the data between 10 and 15 T

Similarly, as described above for the other 2D materials, a characteristic angular dependence of the oscillation amplitude was observed. In first attempts to explain this behavior a warped FS topology was proposed [170, 344, 348]. However, no signs of FS corrugation were found by AMRO experiments [344], proving that the observed angular amplitude change is not caused by this mechanism. Instead, the harmonic-ratio analysis using (4.4) together with the knowledge of the $1/\cos\Theta$ dependence of the effective mass (see Fig. 4.6) can quantitatively describe the angular dependence of the dHvA harmonic ratio as shown in Fig. 4.31. The scatter of the data points is somewhat larger than for the harmonic ratio of α-$(ET)_2NH_4Hg(SCN)_4$ shown in Fig. 4.5 due to the larger effective cyclotron mass ($m_{c0} \approx 3.3\,m_e$) in the former salt. For up to $\pm 12°$ in the Fourier transformed data no second harmonic was detectable (see, e. g., the FFT for $\Theta = 0$ in Fig. 4.30). Therefore, the harmonic ratio was estimated using the noise level of the FFT as an upper limit. At $\Theta = 41.8°$ and $\Theta = 54.8°$ clear minima in the fundamental amplitude and at $\Theta \approx 36°$ a minimum of the second harmonic were found. This behavior can be best described with $\mu_b = 2.6/\cos\Theta$ and a g value close to 2 which is slightly (5%) decreasing with increasing field [279]. The solid line in Fig. 4.31 was obtained with the well-known experimental parameters $T_D = 0.5\,K$, $T = 0.5\,K$, and $B = 14\,T$. From the independently determined values of m_c and the ESR value $g = 2$ an electron–phonon coupling parameter $\lambda \approx 0.27$ can be extracted. The estimation of T_c using the BCS formula (2.6) results in $T_c \approx 6\,K$ with $\Theta_D = 215\,K$ from specific-heat experiments [211, 212]. This is in the right range of the measured $T_c \approx 9.1\,K$ (value obtained by magnetization measurements) and accounts qualitatively for the higher T_c in κ-$(ET)_2Cu(NCS)_2$ compared to α-$(ET)_2NH_4Hg(SCN)_4$.

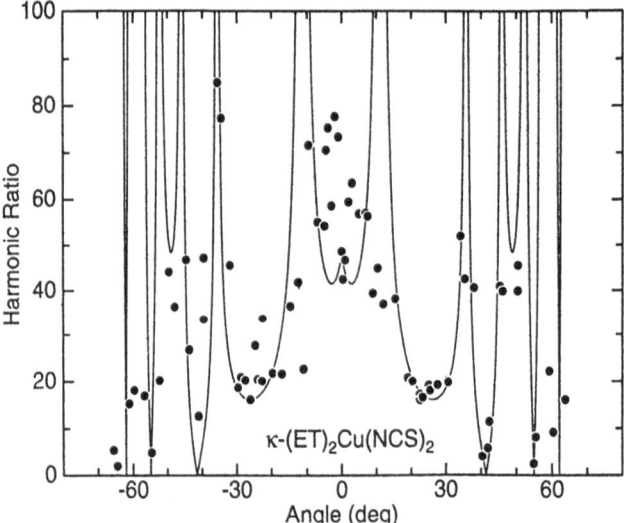

Fig. 4.31. Harmonic ratio of the fundamental to the second-harmonic amplitude of the measured dHvA oscillations for κ-$(ET)_2Cu(NCS)_2$. The solid line represents the calculated ratio using (4.4)

From reflectance experiments of κ-$(ET)_2Cu(NCS)_2$ measuring the far-infrared (wavelength $1223\,\mu m$) transmission in a magnetic field a cyclotron mass of $m_{CR} \approx 1.2\,m_e$ was extracted [282]. This result is similar to the one already discussed for the α-phase salts where m_{CR} also was found to be much smaller than the effective mass m_c extracted from dHvA data. This again has been interpreted as an indication of strong electron–electron interaction in the low-dimensional Fermi liquid. In a Hubbard model an effective on-site Coulomb correlation energy U_{eff} of approximately $0.5\,eV$ for an average energy of the electrons in a half-filled upper band of approximately $60\,meV$ has been estimated [282]. However, a further experimental verification or theoretical justification of this result is missing.

As was mentioned in Sect. 2.3.3, κ-$(ET)_2Cu(NCS)_2$ shows a rapid decrease of the superconducting transition temperature with pressure (-3 K/kbar). Recent SdH experiments under pressure revealed a correlated decrease of the effective cyclotron mass [349]. Up to the critical pressure $P_c \approx 5$ kbar where superconductivity is no longer found the effective mass rapidly falls down to $m_{c0} \approx 2\, m_e$. Further increase of pressure reduced m_{c0} at a much lower rate down to $(1.4 \pm 0.1)\, m_e$ at 16.3 kbar. This suggests that the enhanced effective mass and T_c are directly connected at least in this organic superconductor. The observed strong decrease of m_{c0} cannot merely be an effect of a reduced electron–phonon coupling. Additional quasiparticle interactions seem to play an important role. However, the nature of these interactions and the possible influence on the superconducting properties is an open question and remains to be clarified in future studies.

In some SdH experiments at fields above ~ 20 T in addition to the fundamental frequency, F_0, oscillations of another higher frequency at ~ 3900 T appeared [345, 347]. This corresponds to a cross-sectional area of approximately the whole Brillouin zone. From the calculated band structure (Fig. 2.19) it can be seen that an orbit of this size is only possible by the occurrence of magnetic breakdown (MB), where at the Brillouin zone edge the charge carriers tunnel from the closed hole-like orbit around Z to the open electron-like sheets parallel k_z. Analysis of the SdH data at $\Theta = 0°$ using (4.1) and (4.2) resulted in an energy gap $\epsilon_g = (54 \pm 10)$ K and a magnetic breakdown field $B_{MB} \approx 16$ T. The Bragg reflection angle was set to $2\Theta_{MB} = \pi/2$.[9] The effective cyclotron mass of the large MB orbit was reported to be $m_c^{MB} = 7\, m_e$ [350, 351].

Further dHvA studies revealed an unusual angular dependence of the MB field [279]. In the angular range between 20° and 30° the MB frequency was observed for fields below 15 T, the highest field in this experiment. A typical result in this angular range ($\Theta = -22.7°$) is shown in Fig. 4.32 where the MB frequency is clearly visible as small wiggles on the dominant slow oscillation of $F \approx 650$ T. The inset shows the FFT of the data resolving the MB peak at $F_{MB} \approx 4250$ T besides the slow fundamental F and the second harmonic at $2F$. The MB frequency follows the $1/\cos\Theta$ dependence which has been verified in a recent dHvA experiment [351].

The angular dependent shift of the MB field may have different reasons. One possibility is a k-dependent energy gap, that means there would be a small corrugation of at least one part of the FS. AMRO experiments could not unambiguously resolve the structures verifying the existence of a warped 2D cylinder. Instead angular dependent magnetoresistance minima characteristics of 1D FS sheets have been identified which could be attributed to the open FS running perpendicular to the k_b plane (see Fig. 2.19a) [352]. However, a possible additional warping of the 2D FS could not be excluded. Another possible reason for the strong MB signal around $\sim 25°$ might be the usual angular dependence of the effective mass. With the measured values $m_c^{MB} \approx 6.9\, m_e$ and $\lambda \approx 0.27$ the bare band mass is close to $5.5\, m_e$ for

---

[9] In a recent thorough dHvA study $\epsilon_g = 73$ K and $B_{MB} = 30.5$ T is reported [351].

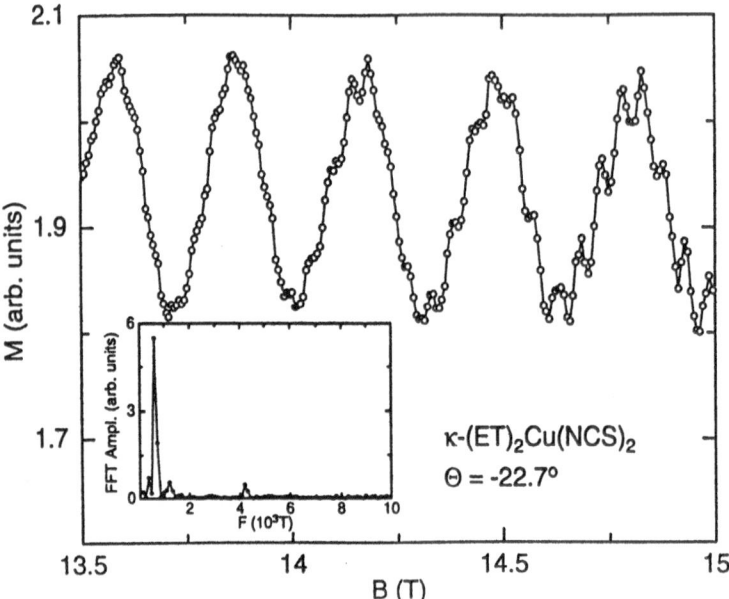

Fig. 4.32. dHvA signal of κ-(ET)$_2$Cu(NCS)$_2$ at $\Theta = -22.7°$ and $T \approx 0.5\,\mathrm{K}$ showing the appearance of a higher frequency F_{MB}. The inset shows the FFT of the data resolving the MB peak at $F_{\mathrm{MB}} \approx 4250\,\mathrm{T}$

$\Theta = 0°$. Therefore, the amplitude of the MB signal around this angle might be largely reduced due to a near spin-splitting zero for this orbit. Indeed, in a dHvA experiment carried out in fields up to $\sim 30\,\mathrm{T}$ spin-splitting zeros for the MB oscillations could also be observed at $\Theta = 10°, 35°$, and $45°$ [351]. This is within error bars consistent with $\lambda \approx 0.27$.[10]

Due to the relatively large upper critical field and the availability of high-quality crystals of κ-(ET)$_2$Cu(NCS)$_2$ it was also possible to observe dHvA oscillations in the superconducting vortex state [32]. This has also been possible so far for a few other inorganic type II superconductors such as NbSe$_2$ [353], V$_3$Si [354], Nb$_3$Sn [355], and recently for YNi$_2$B$_2$C [356, 357]. Figure 4.33 shows the dHvA result for κ-(ET)$_2$Cu(NCS)$_2$ at $T = 20\,\mathrm{mK}$ and $\Theta = 29°$ together with the behavior expected from the Lifshitz–Kosevich formula (3.6) for a constant Dingle temperature [32]. For higher fields above B_{c2} the experimental data can reasonably be described by (3.6). Below B_{c2} the dHvA signal is attenuated more quickly than predicted. This behavior has also been found for the dHvA signal in the vortex state of the other above-mentioned superconductors. For these materials it was sometimes possible to observe dHvA oscillations well below B_{c2}. In most theories the dHvA signal in the superconducting state is thought to be entirely due to the normal state quasiparticles

---

[10] The authors in [351] neglected a possible mass-enhancement due to electron–phonon interaction and attributed the spin-splitting zeros to a g value of ~ 1.55.

which travel in the superconductor just below B_{c2} in a slightly inhomogeneous field. This effect would be experimentally equivalent to an effective increase in Dingle temperature. A theoretical interpretation of the damping is based on the assumption of an enhanced broadening of the Landau levels due to an additional scattering of the quasiparticles in the vortex state. This would result in a field dependent scattering rate $\tau^{-1} = \tau_0^{-1} + \tau_s^{-1}(B)$ [358]. Although this model should only be valid close to B_{c2} it could successfully account for the experimentally observed behavior in the inorganic superconductors [353, 354, 355]. In contrast, in a different theory the oscillations in the superconducting state are attributed to oscillations of the superconducting ground state itself [359]. This model also predicts an additional damping due to the presence of the energy gap in the superconducting state. The field dependent damping of the dHvA data in YNi_2B_2C can be described by this model very well down to below $0.5\,B_{c2}$ [357]. In the present case for $\kappa\text{-}(ET)_2Cu(NCS)_2$ the model of [358] was used. Due to the fact that the precise value of B_{c2} is unknown (see Sect. 2.3.3), B_{c2} was used as a fit parameter to describe the field dependent change of the Dingle temperature, respectively the scattering rate. With the BCS value for the superconducting gap, $\Delta \approx 1.6\,\text{meV}$, and a spread of B_{c2} values an average B_{c2} of $4.6\,\text{T}$ was estimated [32]. This is much less than the value extrapolated from magnetization measurements (see Table 2.2) [189]. Additional work has to be done to clarify this discrepancy.

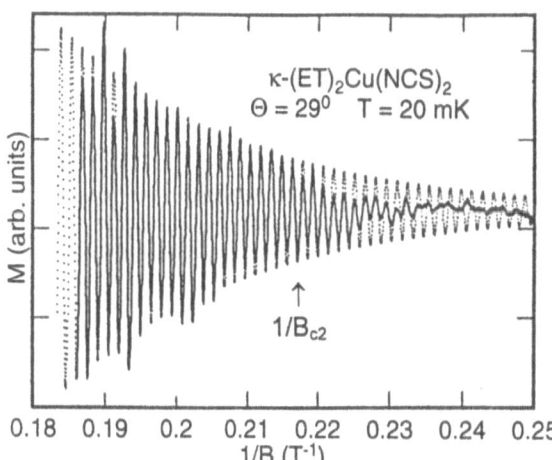

Fig. 4.33. dHvA signal of $\kappa\text{-}(ET)_2Cu(NCS)_2$ at $T = 20\,\text{mK}$ and $\Theta = 29°$ close to the upper critical field (solid line). The slowly varying background is subtracted. The dotted line is the signal predicted by (3.6) using the parameters derived from data taken well above B_{c2}. The arrow indicates the fit parameter B_{c2}. From [32]

The gross features of the FS of $\kappa\text{-}(ET)_2Cu(NCS)_2$ have been verified independently by a measurement of the angular correlation of positron annihilation radiation [360]. The pair momentum distribution was compared to the 2D tight-binding band-structure calculation [29]. The closed orbit was estimated to occupy $(13 \pm 1)\%$ of the first Brillouin zone compared to 15% obtained from the dHvA results. In addition, the open sheets running parallel to YM (see Fig. 2.19a) were discovered to be in agreement with the SdH and dHvA measurements.

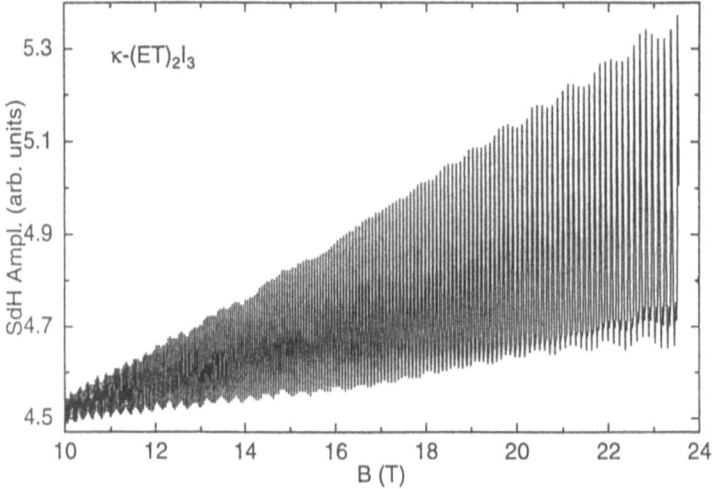

Fig. 4.34. SdH signal of κ-$(ET)_2I_3$ at $T \approx 0.5$ K for a field applied perpendicular to the ET planes. From [363]

Another well investigated κ-phase salt is κ-$(ET)_2I_3$. In the first dHvA experiment two frequencies of $\sim 100\%$ and $\sim 20\%$ of the first Brillouin zone were reported [361]. For the larger orbit an effective cyclotron mass of $m_c \approx 2.7\,m_e$ was obtained. Later work on dHvA and SdH effect extended this first result to higher fields and examined the angular dependence of the FS extremal areas [192, 362, 363]. An example of the SdH data is shown in Fig. 4.34 for a large field range [363]. Clearly the low and high frequencies are visible. Both frequencies were found to follow the $1/\cos\Theta$ dependence with minimum values of ~ 570 T and ~ 3880 T for the field applied perpendicular to the highly conducting planes. The result is consistent with the predicted FS for κ-$(ET)_2I_3$ (Fig. 2.19b) [147, 163]. The overall shape of the FS is almost like the one for κ-$(ET)_2Cu(NCS)_2$ except for the value of the gap between the two bands at the Fermi energy. For κ-$(ET)_2I_3$ the gap is very small and, therefore, the MB frequency appears at very low fields.

Some discrepancies concerning the effective cyclotron mass were found. In contrast to the first published result, all later experiments consistently report a value of $m_c = (3.9 \pm 0.1)\,m_e$ for the large orbit at $\Theta = 0°$ and fields below ~ 13 T (see also Fig. 4.6) [192, 362, 363]. The effective mass for

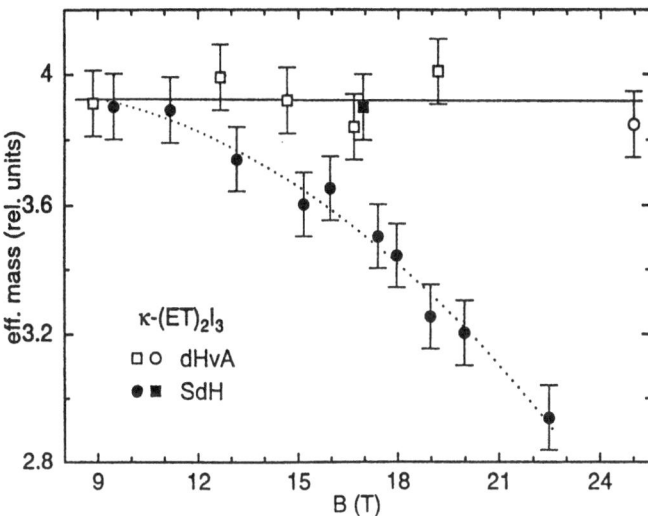

Fig. 4.35. Field dependence of the effective cyclotron mass of κ-(ET)$_2$I$_3$. The closed circles are obtained from SdH data for B perpendicular to the ET layers. The open symbols are from dHvA data, the closed rectangle from SdH data with a slightly canted field. The solid line shows the field independent mass. The dotted line is a guide to the eye. From [363]

the small orbit is $m_c^{\text{small}} = (1.9 \pm 0.1) \, m_e$ [364]. An unusual behavior of the field dependence of the large-orbit effective mass m_c was found. Figure 4.35 shows the values of m_c extracted from the temperature dependence of the SdH and dHvA amplitudes for different fields (see (3.7) in Sect. 3.1.1) [363]. Above approximately 13 T, m_c obtained from the SdH data at $\Theta = 0°$ decreases from $\sim 3.9 \, m_e$ to $\sim 2.9 \, m_e$ at 23 T. This decrease, however, is only observed when the field is applied perpendicular to the conducting planes. The SdH and dHvA values extracted for inclined fields are consistent with $m_c = 3.9 \, m_e / \cos \Theta$. The dHvA data were measured by the torque method. Therefore, the sample could not be investigated at $\Theta = 0°$.

The origin for this apparent field dependence of m_c is not fully understood. However, the nearly perfect two-dimensionality of κ-(ET)$_2$I$_3$ discussed in Sect. 2.3.3 seems to be the principal reason for the observed strange temperature and field dependence of the magnetic quantum oscillations. It was suggested that in this extremely 2D system quasiparticles with fractional statistics [365] may occur if the cyclotron orbits lie within an individual 2D conducting plane [363]. Since these quasiparticles do not obey Fermi statistics they should not contribute to the quantum oscillations observed. Hence, the effective cyclotron mass determined by the 3D Lifshitz–Kosevich formula could be underestimated. Further experimental verification for this suggestion is lacking.

For lower fields the dHvA oscillations do not behave as predicted by the Lifshitz–Kosevich formula. Figure 4.36 shows typical dHvA oscillations of

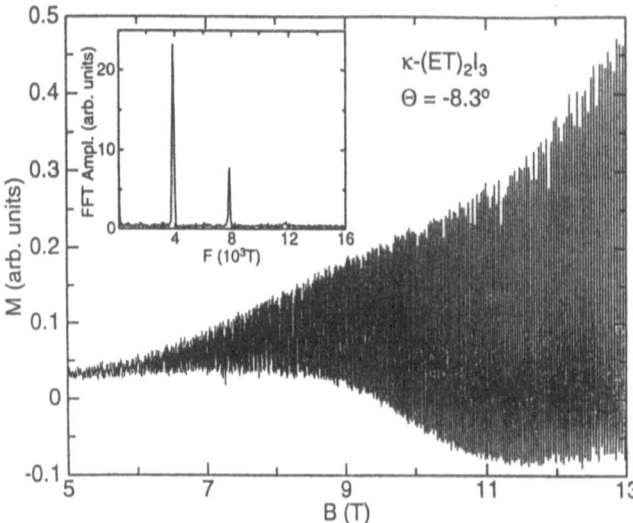

Fig. 4.36. dHvA signal of κ-(ET)$_2$I$_3$ at $T \approx 0.5$ K. The inset shows the FFT of the data between 11 and 13 T

κ-(ET)$_2$I$_3$ at $T \approx 0.5$ K up to 13 T. In the inset the Fourier transform of the data between 11 and 13 T is plotted. Besides the fundamental dHvA frequency at $F \approx 3900$ T a strong second harmonic at $2F$ and even a small third harmonic at $3F$ are visible. This strong harmonic content is not in accordance with the expected behavior for the experimentally obtained parameters and the 3D Lifshitz–Kosevich formula. This could be related to the above-mentioned strong harmonic content observed in β_H-(ET)$_2$I$_3$ at angles where only one extremal FS area exists (see Fig. 4.18).

For κ-(ET)$_2$I$_3$ no beating of the SdH or dHvA signal was observed even for the lowest fields where oscillations were seen. This proves the extremely 2D character of this organic superconductor. As an upper limit for the transfer integral t the extraordinary small value of $t/\epsilon_F < 1/5000$ can be estimated. This almost perfectly 2D electronic structure might be the reason for the unusual behavior of the SdH and dHvA oscillations, especially for the field dependence of the effective mass at higher fields where the 3D Lifshitz–Kosevich theory no longer works.

The angles where spin-splitting zeros could be observed in κ-(ET)$_2$I$_3$ have been shown in Fig. 4.22. The zeros follow the $1/\cos\Theta$ dependence perfectly, yielding a value of $g\mu_b = 8.64$. As discussed above for β_H-(ET)$_2$I$_3$ and β-(ET)$_2$IBr$_2$, this means that either the g value must be considerably larger than 2 or the determination of the effective cyclotron mass is not correct. Further investigations are necessary to obtain a better understanding of the influence of a perfect 2D FS on magnetic quantum oscillations.

As mentioned recent experiments under pressure succeeded in the observation of SdH oscillations in the isostructural "high-T_c" compounds

κ-(ET)$_2$Cu[N(CN)$_2$]Br and κ-(ET)$_2$Cu[N(CN)$_2$]Cl [341]. In the former salt at the highest applied pressure of ~ 9 kbar weak resistivity oscillations with $F_0 \approx 156$ T were observed. The usual $1/\cos\Theta$ dependence was found. An effective cyclotron mass of $m_c \approx 0.95\, m_e$ with $T_D \approx 3.5$ K was reported [341]. In a different experiment extending the measurements to higher fields an additional frequency with $F_{MB} \approx 3800$ T accounting for a breakdown orbit was found. These results suggest that the closed part of the FS around Z is reduced in size by approximately a factor of 4 compared to κ-(ET)$_2$I$_3$. In addition, the gap between the closed and open parts of the FS is clearly larger than expected from band-structure calculation. Why the SdH effect is only observable above the appreciable pressure of ~ 9 kbar and the reason for the discrepancy between the experimentally obtained and the expected FS is unclear. As a possible origin for the absent SdH effect under ambient conditions a low-temperature magnetic ordering with an inhomogenous local magnetic field within the sample is suggested [341]. This is based on some experimental evidence from a vibrating-reed study that a field-induced magnetic phase transition occurs in κ-(ET)$_2$Cu[N(CN)$_2$]Br [366].

For κ-(ET)$_2$Cu[N(CN)$_2$]Cl, whose peculiar superconducting properties and rich low-pressure phase diagram was discussed in Sect. 2.3.3, SdH oscillations were observed above ~ 3.8 kbar [341]. At this pressure both low and high frequency oscillations were observed. The extracted extremal areas of the FS are consistent with the predictions for the κ phase. At ~ 6 kbar a beating of the MB frequency was observed which suggests an appreciable warping of the FS. Further experiments are needed to obtain a better knowledge on the band structure of these organic "high-T_c" superconductors.

For two other κ-phase substances SdH and dHvA signals have been reported. The first material is κ-(ET)$_2$Ag(CN)$_2$H$_2$O with $T_c \approx 5$ K [367]. A fundamental dHvA frequency "nearly the same as in κ-(ET)$_2$Cu(NCS)$_2$" with no explicitly stated value has been reported [368]. This means presumably that for the field perpendicular to the conducting planes a frequency of ~ 600 T has been observed. The dHvA signal shows a clear beating behavior which is explained by a warped FS due to a considerably stronger interlayer overlap integral than in the other κ-phase salts. The reported effective cyclotron mass is $m_c = 2.7\, m_e$ [368]. This value seems to be somewhat small compared to the other isostructural salts. No MB frequency was seen in the experimentally accessible field range up to ~ 12 T. Since κ-(ET)$_2$Ag(CN)$_2$H$_2$O is structurally equivalent to κ-(ET)$_2$Cu(NCS)$_2$, a gap between the 2D and 1D FS and a similar MB field is also expected for this material. So far neither verification of this result nor a systematic angular dependent study has been reported for κ-(ET)$_2$Ag(CN)$_2$H$_2$O.

The second material is κ-(DMET)$_2$AuBr$_2$ with $T_c \approx 1.9$ K [17]. This is the first charge transfer salt with a donor molecule other than ET where the dHvA effect was observed [369]. DMET (= dimethyl-ethylenedithio-diselenedithiafulvalene) is an asymmetric hybrid molecule composed of one half TMTSF and one half ET as shown in Fig. 1.1 (page 3). Not very much information has been extracted from the only dHvA study so far [369]. The oscillating signal is very similar to the behavior seen in κ-(ET)$_2$I$_3$. At relatively low fields (~ 9 T) two frequencies of ~ 4500 T and ~ 920 T corresponding to slightly above 100% and ~ 21% of the first Brillouin zone were found. This is consistent with the predicted nearly gapless band structure for the κ phase with space group $P2_1/a$. The effective cyclotron masses reported were $3.8\,m_e$ and $6.0\,m_e$ for the small and large orbit, respectively. These values are close to the masses measured for κ-(ET)$_2$Cu(NCS)$_2$ but considerably larger than in κ-(ET)$_2$I$_3$. The wave form of the dHvA signal was found to be saw-tooth like which was attributed to the 2D character of the FS [369]. However, the shape of the saw-tooth is not as predicted for a 2D electronic structure [326]. A verification of the present results with a systematic angular dependent study would be needed to clarify the many open questions for κ-(DMET)$_2$AuBr$_2$.

In summary, the experimentally obtained FS topology of the κ phase is principally in good agreement with the predicted band structure. Some of the salts exhibit an extreme 2D electronic structure, even by standards of ET compounds. Especially for κ-(ET)$_2$I$_3$ strong deviations from the 3D Lifshitz–Kosevich theory are observed. On the other hand, the dHvA signal of κ-(ET)$_2$Cu(NCS)$_2$ is fully compatible with the 3D theory and even the extraction of the electron–phonon coupling constant was possible. The effective cyclotron masses for the MB orbit are $\sim 3.9\,m_e$ for the first and $\sim 7\,m_e$ for the latter salt. How far this reflects the different T_cs of these two compounds needs further investigation.

4.2.4 Θ Phase

Another modification of the compound (ET)$_2$I$_3$ crystallizes in the so-called Θ-type structure [147]. Strictly speaking, however, the exact composition of the salt should be written as (ET)$_2$(I$_3$)$_{1-x}$(AuI$_2$)$_x$ with $x < 0.02$. This arises because the crystals grow from a solution containing ET, (n-C$_4$H$_9$)$_4$I$_3$, and a small amount of (n-C$_4$H$_9$)$_4$AuI$_2$. The crystal structure initially gave rise to some controversy. In first investigations an orthorhombic structure was determined [147]. Later detailed inspection of the I$_3$ molecule X-ray reflections revealed a superstructure with an approximately double-sized monoclinic unit cell (see Sect. 2.3.1 and Tab. 2.1) [149, 370]. Consequently, the calculated band structure for the different unit cells differs. Figures 4.37(a) and (b) show the FSs calculated for the orthorhombic and the monoclinic structure, respectively. In both cases the FS topology is close to a free-electron like picture. However, because of the larger monoclinic unit cell the 2D free-electron Fermi circle cuts the smaller Brillouin zone at more points resulting in three instead of one closed portion of the FS.

In this somewhat unsatisfactory situation dHvA, SdH, and AMRO experiments were successful in resolving the correct FS topology which consequently were found to have neither of the proposed shapes. The supercon-

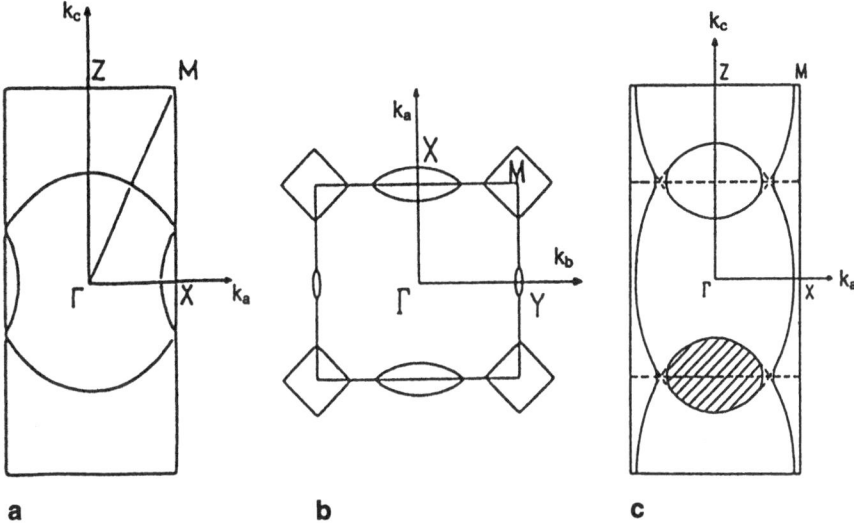

Fig. 4.37. Calculated FS of Θ-(ET)$_2$I$_3$ for (a) the orthorhombic structure (from [147]), (b) the monoclinic structure (from [149]), and (c) the proposed FS consistent with the observed magnetic quantum oscillations (from [372])

ducting properties of Θ-(ET)$_2$I$_3$ are very sample dependent. In fact, not all samples showed a superconducting transition. For some the resistivity was zero at $T_c = 3.6$ K, for others only a partial or even no resistivity decrease was observed [147]. Magnetization measurements also showed this unsystematic behavior. In some crystals bulk superconductivity was verified by large diamagnetic shielding signals while others showed only approximately 1/1000 of this signal [371]. The reason for this strong sample dependence is unclear. It was assumed that crystal imperfections might decisively influence the superconducting properties [371]. However, the first dHvA results reported were done on a non-superconducting sample [298, 372]. Therefore, a good crystal quality, i.e., a small Dingle temperature, is not the dominant factor for superconductivity in Θ-(ET)$_2$I$_3$.

Figure 4.38 shows the measured magnetization at $T = 0.5$ K for a field applied perpendicular to the conducting planes [372]. Clearly two different dHvA frequencies are visible in the signal. A slow oscillation of ~ 730 T is superimposed on a dominant high frequency of ~ 4170 T.[11] These frequencies correspond to FS areas which are more consistent with the band-structure calculation for the orthorhombic unit cell (Fig. 4.37a) but are in disagreement with the predicted band structure for the monoclinic cell (Fig. 4.37b).

From Drude-like reflectance spectra at $T = 16$ K a free-electron like elliptical FS area with no crossing of the orthorhombic Brillouin zone was inferred

---

[11] In a later systematic investigation of the angular dependence of the SdH signal at $\Theta = 0°$ frequencies of 779 T and 4234 T were reported [373].

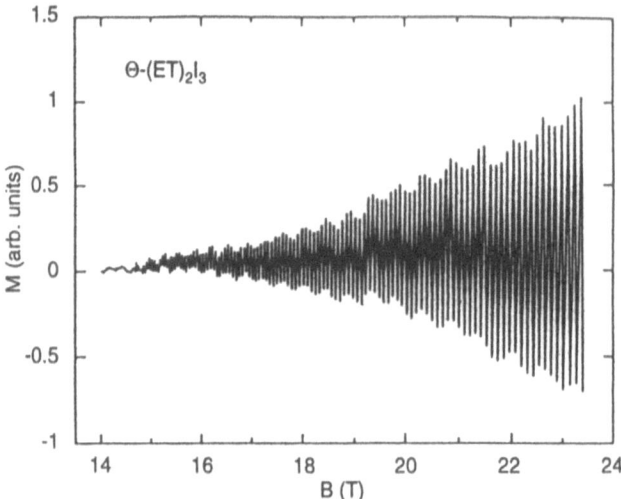

Fig. 4.38. dHvA signal of Θ-(ET)$_2$I$_3$ at $T = 0.5$ K for a field applied perpendicular to the ET planes. From [372]

[374]. However, since the superstructure of the I$_3$ anions doubles the crystal lattice vector along c (which becomes a in the monoclinic nomenclature) a SDW-like reconstruction of the orthorhombic band structure was proposed [298, 372]. In a simple picture the Brillouin zone of Fig. 4.37a should be folded at the midpoint of the ΓZ line. With the assumption based on the reflectance data of an elliptical FS elongated along ΓZ the folding results in the FS shown in Fig. 4.37c. Indeed, this FS agrees best with all experimental data where the shaded area corresponds to the slow dHvA oscillations. The high dHvA frequency results from a MB orbit across the small gap indicated by the dashed line. The effective cyclotron masses were determined in a later experiment and are $3.6\,m_e$ and $2.0\,m_e$ for the large and small orbit, respectively [375].

Successive high-resolution AMRO experiments shown in Fig. 4.39 verified the proposed FS in an impressive way [376]. As mentioned in Sect. 3.3, Θ-(ET)$_2$I$_3$ was one of the first compounds where AMRO, i. e., resistance oscillations periodic in $\tan\Theta$, were observed [258]. These results, which were reproduced later [377], are understood by the warped FS model explained above. The period of the oscillations is related to the Fermi wave vector via (3.18). In the experimental data shown in Fig. 4.39 not only the previously reported fast AMROs but also slow ones (indicated by small dashes) were observed [376]. The insets of Fig. 4.39 show the peak numbers of the (a) fast and (b) slow oscillation frequency vs $\tan\Theta$. From the slopes for different field rotation planes $k_F(\Theta)$ could be constructed. The resulting two ellipsoidal FSs are in good agreement with the proposed topology of Fig. 4.37c with respect to both form and area.

An interesting point of this result is the fact that obviously the FS of Θ-(ET)$_2$I$_3$ is corrugated due to a not small but finite overlap between the conducting layers. However, in both dHvA and SdH signals no signs of a beating behavior as described, e. g., for β-(ET)$_2$IBr$_2$ were found.[12] Therefore, since oscillations were observed around 10 T the transfer integral t must be very small. This, on the other hand, means that the AMRO are already highly sensitive to a small corrugation.

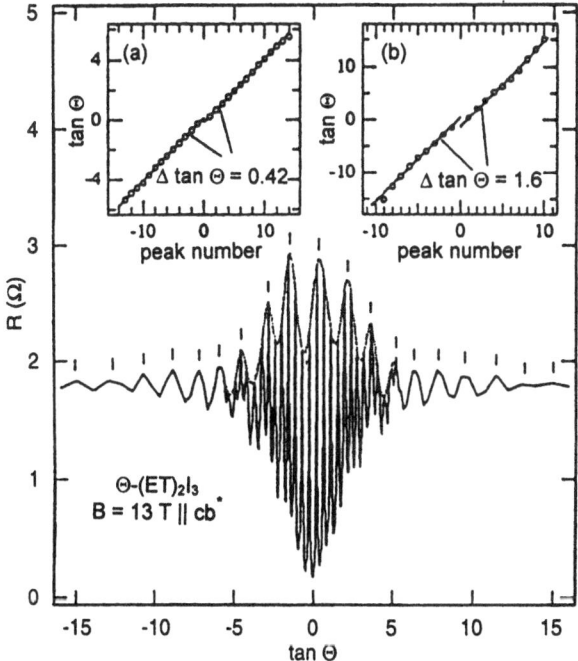

Fig. 4.39. The main panel shows the angular dependence of the resistance of Θ-(ET)$_2$I$_3$ in $B = 13$ T for field rotation in the cb^* plane (orthorhombic notation). Two superimposed AMROs are visible. The insets show the peaks of (a) the fast and (b) the slow oscillation vs $\tan\Theta$. From [376]

The highly 2D nature of the FS may have directly been seen in a saw-tooth form of the dHvA signal [298, 372]. The shape of the saw-tooth described in this experiment is consistent with the predicted form of dHvA oscillations in a strictly 2D metal [326]. However, a more explicit analysis of the data corresponding to the proposed 2D formula has not been done. In addition, all later experiments were analyzed with the usual 3D Lifshitz–Kosevich formula. In the previously discussed measurements of other salts where an enhanced harmonic content of the magnetic quantum oscillations were reported such a

---

[12] The nodes observed in one experiment [375] were attributed to twinning of the sample investigated.

good resolved saw-tooth form of the signal as in Θ-$(ET)_2I_3$ could never be observed. The detailed experimental manifestation of 2D dHvA oscillations is still far from being unequivocal and remains to be investigated further.

As a final point, in Θ-$(ET)_2I_3$ at fields below $\sim 2\,T$ additional very slow SdH oscillations were observed [373]. The reported frequencies were somewhere between 2 and 12 T and were more 3D in character. Very small effective cyclotron masses of $0.05\,m_e$ and $0.014\,m_e$ were obtained. From the predicted band structures these extremely small 3D FS pockets cannot be understood. Further work is necessary to verify these observations.

So far no other Θ-phase material has been found to show magnetic quantum oscillations. Therefore, in contrast to the other phases discussed relations between the various physical properties of different materials and the electronic band structure cannot be investigated.

4.2.5 Other Materials

Besides the organic metals and superconductors based on the donor molecule ET numerous metallic organic charge transfer salts composed of different donors and acceptors have been synthesized. However, only a small number of these salts become superconducting, usually at rather low temperatures, and even less have been investigated in the same detail as the ET compounds. Apparently, the generally low crystal quality of the non-ET materials permitted only a few Fermi surface studies by SdH or dHvA experiments.

The organic donor molecule BEDO–TTF (see Fig. 1.1) which is closely related to ET was originally synthesized with the aim of reducing the molar weight of the charge transfer salts. It was hoped to obtain ET-analogous compounds with increased T_c. However, to date only two superconducting salts, $(BEDO–TTF)_3Cu_2(NCS)_3$ [20] and $(BEDO–TTF)_2ReO_4(H_2O)$ [21], both with T_cs around $1\,K$ have been synthesized. A possible reason for this might be the rather poor crystal quality and disorder effects in the anion layers. The latter effect was found in $(BEDO–TTF)_2ReO_4(H_2O)$ [21]. The crystal structure at room temperature is monoclinic with space group $P2_1/a$, $a = 12.16\text{Å}$, $b = 34.05\text{Å}$, $c = 8.091\text{Å}$, $\beta = 123.44°$, $V = 2795.5\text{Å}^3$, and $Z = 4$. This charge transfer salt shows several phase transitions upon cooling from room temperature as manifested in its transport properties. At $\sim 210\,K$ a metal–metal transition and at $\sim 35\,K$ a change from metallic to semiconducting behavior was observed. Finally, around $2.5\,K$ the resistivity starts to decrease strongly which was associated with the onset of superconductivity. However, ac-susceptibility measurements show that only around $\sim 0.9\,K$ does a diamagnetic signal evolve and the transition is complete at $\sim 0.05\,K$ [21]. With the application of the moderate pressure of $\sim 1\,kbar$ the resistivity increase below $35\,K$ can be suppressed and the superconducting transition is sharpened [378].

SdH measurements unraveled both the band structure of $(BEDO–TTF)_2$-$ReO_4(H_2O)$ and resolved the question of a DW transition in the ambi-

ent pressure state. A competition between a SDW phase and superconductivity is thought to be responsible for the low-temperature behavior observed in this salt [379]. Figure 4.40 shows the magnetoresistance of (BEDO–
TTF)$_2$ReO$_4$(H$_2$O) for different temperatures and for fields applied perpendicular to the planes [379]. The Fourier transformation of the data at $T = 0.5$ K
resolves two SdH frequencies present in the measured signal. The frequencies
are very low with $F_1 = (37 \pm 3)$ T and $F_2 = (75 \pm 0.3)$ T corresponding to
$\sim 0.7\%$ and $\sim 1.5\%$ of the first Brillouin zone cross section. The effective cyclotron masses of the orbits are relatively large with $m_{c1} = (1.15 \pm 0.1) m_e$
and $m_{c2} = (0.9 \pm 0.05) m_e$, respectively. The larger effective mass for the
smaller orbit is highly unusual and might hint at many-body interactions re-

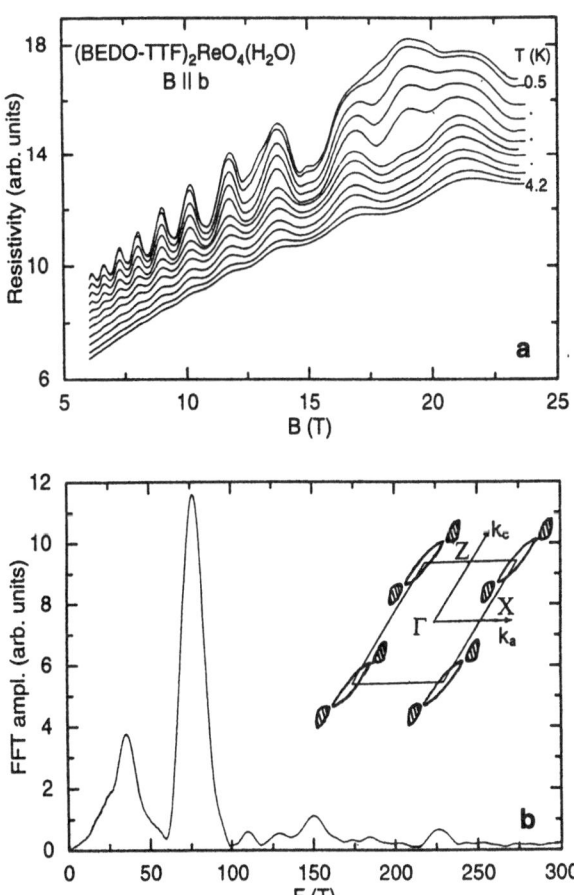

Fig. 4.40. (a) Magnetoresistance of (BEDO–TTF)$_2$ReO$_4$(H$_2$O) for different temperatures between 0.5 and 4.2 K. (b) Fourier transformation of the data at
$T = 0.5$ K. The inset shows the calculated tight-binding 2D FS. From [379]

sponsible for the mass enhancement. Both SdH frequencies were measured up to nearly 80° and showed the expected 2D $1/\cos\Theta$ behavior [379] perfectly.

The fact that $F_2 \approx 2F_1$ might suggest that both frequencies are due to the same orbit. However, the amplitude of the oscillation corresponding to F_2 is always larger than that of F_1. Tight-binding band-structure calculations with the room temperature lattice parameters resulted in the 2D FS shown in the inset of Fig. 4.40b [379]. A hole pocket centered around M and two essentially equal-area electron pockets (shaded) with sizes of $\sim 3.4\%$ and $\sim 1.7\%$, respectively, are obtained. The fact that the experimentally extracted areas are of approximately half these sizes is attributed to changed lattice parameters at temperatures below the phase transitions. Low-temperature X-ray measurements or SdH experiments under pressure to confirm this idea are missing. Closer examination of the FS reveals a so-called hidden nesting, which means that different parts of the FS can be brought on top of each other by a nesting vector $\boldsymbol{Q}$. In the present case this vector is $\boldsymbol{Q} = (a^*, 0)$. The nesting condition is thought to initiate a SDW transition around 35 K. However, the imperfect nesting and possibly "internal pressure" induced by thermal contraction may result in the reentrance of the original metallic phase at lower temperatures [379].

In conclusion, the SdH data are in satisfactory agreement with the band-structure calculations. The extracted electronic structure was used as a basis for the understanding of the transport properties observed. The relatively low superconducting transition temperature of $(BEDO\text{--}TTF)_2ReO_4(H_2O)$ does not seem to be due to the anion disorder observed. The Dingle temperature was estimated to $T_D \approx 1.4$ K. The FS studies rather suggest that the hidden nesting prevents a higher T_c. Additional information on the electronic structure for other BEDO–TTF compounds would also be helpful to understand the unexpectedly different superconducting and electronic properties compared to the ET salts better.

Another family of organic metals and superconductors is based on the acceptor molecule $M(dmit)_2$ with $M = $ Ni, Pd, and Pt. A variety of these salts becomes superconducting at low temperatures [22, 23]. Since the organic anions form the conduction bands the lowest unoccupied molecular orbitals (LUMO) are the relevant ones for band-structure calculations. However, to date there is only little and in addition mostly indirect experimental evidence for the predicted electronic structures reported. Therefore, systematic investigations of the FS by direct SdH or dHvA measurements are needed for verification of the calculations. The only salt where SdH oscillations have been reported is $\alpha\text{-Me}_2Et_2N[Ni(dmit)_2]_2$ where Me_2Et_2N is $(CH_3)_2(C_2H_5)_2N$. This compound is metallic down to 0.5 K but not superconducting. At ~ 245 K a sharp increase of the resistance towards lower temperatures is observed [380]. This is attributed to a first-order phase transition. X-ray measurements revealed that at this temperature the cations $Me_2Et_2N^+$ become ordered due to freezing of rotational degrees of freedom of the ethylene groups around

N–C bonds. Below this phase transition the interlayer periodicity is doubled and the crystal symmetry changes from $C2/c$ to $P2_1/c$ [381].

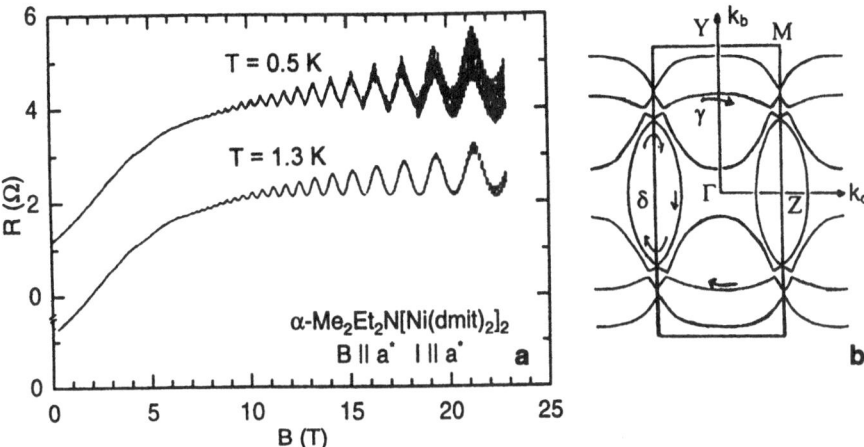

Fig. 4.41. (a) Field dependence of the resistance of α-Me$_2$Et$_2$N[Ni(dmit)$_2$]$_2$ for two temperatures. (b) Calculated FS from lattice parameters at 11 K. From [380]

Figure 4.41a shows the measured resistivity for two different temperatures and fields up to 23 T [380]. The Fourier transformation of the data revealed up to four different SdH frequencies which are summarized in Table 4.2. The two lower frequencies were already observed in preliminary SdH experiments up to 8 T and showed the typical 2D $1/\cos\Theta$ dependence [382]. The SdH frequency labeled δ was found to be only very weak and was present only in a restricted field range. Therefore, it is assumed that the corresponding orbit belongs to a magnetic breakdown orbit. The experimentally extracted effective cyclotron masses are given in Table 4.2 except for the δ orbit where the signal was too weak to determine a reliable value.

Table 4.2. Experimentally obtained FS parameters of α-Me$_2$Et$_2$N[Ni(dmit)$_2$]$_2$. F_i is the SdH frequency ($i = \alpha$, β, γ, δ), a_i/BZ is the corresponding area divided by the area of the first Brillouin zone ($= 4.47 \times 10^{15}$ cm^{-2}) in percent, and m_c/m_e is the effective cyclotron mass in relative units. For the δ orbit no effective mass could be extracted. From [380]

	F_i(T)	a_i/BZ	m_c/m_e
α	10.6	0.2	0.06±0.01
β	214	4.5	0.166±0.01
γ	4021	86	4.3±0.2
δ	520	11	–

The orbital areas obtained from the SdH oscillations disagree with the band-structure calculation based on the room temperature crystal structure [383]. This, however, is not very surprising since the material undergoes the mentioned phase transition at ~ 245 K with a concomitant change of crystal symmetry. Therefore, a band-structure calculation with the measured lattice parameters at 11 K has been made [381]. The resultant FS is shown in Fig. 4.41b. Yet, this calculation cannot reproduce all the observed SdH frequencies, especially the two low ones (α and β). However, the lens-like orbit around Z might be responsible for F_δ. The calculated area of 15% of the first Brillouin zone is close to the experimental value. At higher fields the band structure suggests a possible MB orbit similar to that observed in κ-$(ET)_2Cu(NCS)_2$ (see Sect. 4.2.3). The calculated area of this MB orbit is $\sim 89\%$ which is close to the measured value of the γ orbit. Although this interpretation accounts for the two higher SdH frequencies the origin of the α and β orbit remains unexplained by the present band structure. In X-ray experiments an additional superstructure with lattice constant $2b$ was detected at 11 K [381]. Therefore, a further reconstruction of the FS might occur. For verification of this possibility and the exact determination of the superstructure, however, additional measurements are necessary.

These are the only SdH results to date of a $M(dmit)_2$ salt which show clearly that at least for these materials simple tight-binding calculations have to be done with great care. The determination of the real low-temperature crystal structure is essential. Further work for more $M(dmit)_2$ compounds would be needed to get a better understanding of the electronic properties which could give some hints for the important factors governing the appearance of superconductivity in these systems.

The best-conducting organic salts ever found to date are compounds of the series $(R_1, R_2\text{--DCNQI})_2M$, where $R_1, R_2 = CH_3, CH_3O, Cl, Br$, and I. DCNQI is N,N'-dicyanoquinonediimine, and $M = Cu, Ag, NH_4, Tl$, or alkaline metals [384]. The π-acceptor molecule $R_1, R_2\text{--DCNQI}$ is shown in Fig. 4.42.

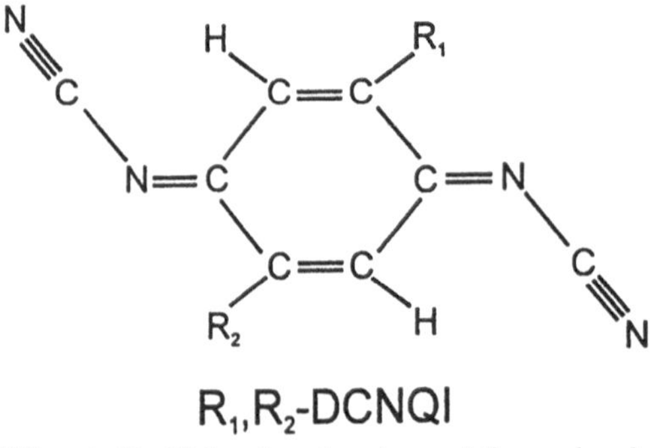

R_1,R_2-DCNQI

Fig. 4.42. Molecular structure of the molecule R_1, R_2–DCNQI

One of the most investigated salts based on this acceptor molecule is (DMe–DCNQI)$_2$Cu, where DMe is dimethyl ($=$(CH$_3$)$_2$). This compound is especially interesting because a metal–insulator transition, sometimes even with a reentrance to metallic behavior, can be induced either by pressure or by the substitution of hydrogen with deuterium in the organic complex [385]. The crystal structure has tetragonal symmetry with the space group $I4_1/a$. The planar DMe–DCNQI anions are stacked in 1D columns along the tetragonal c axis. The LUMO consists of p_π orbitals and forms a wide 1D conduction band which is partly filled with electrons donated by the cations. The Cu ions are in a mixed valence state, namely [Cu^{2+}]:[Cu$^+$] $= 1$:2 [386]. They interconnect four DCNQI molecules in a tetrahedral coordination.

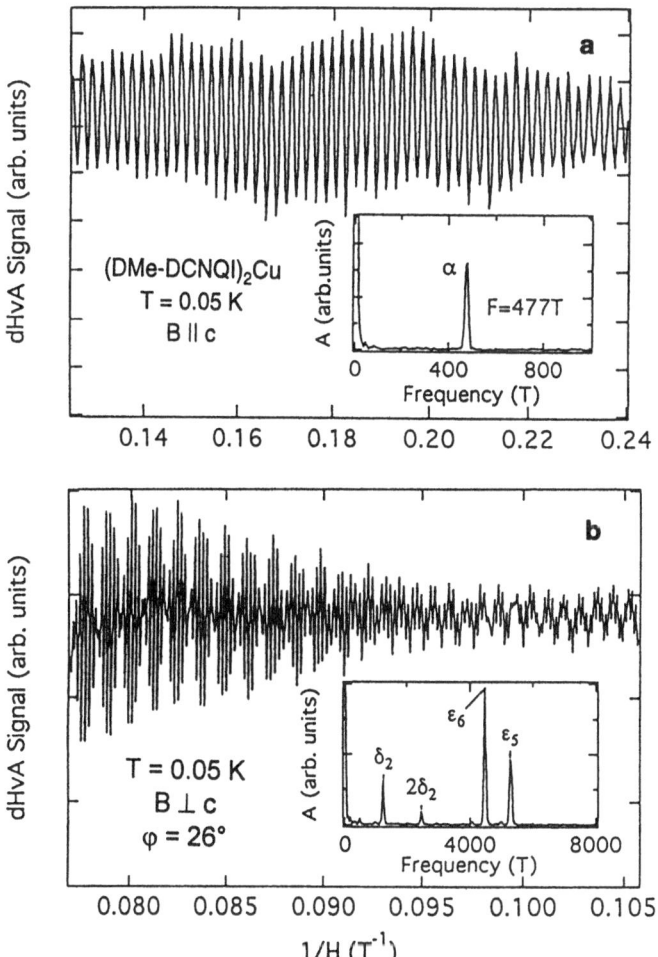

Fig. 4.43. dHvA signal of (DMe–DCNQI)$_2$Cu at $T = 0.05$ K for (a) $B \parallel c$ and (b) the field in the ab plane enclosing an angle of $\varphi = 26°$ with the [110] direction. The insets show the corresponding Fourier transformations. From [388]

Band-structure calculations show that the 3d level of the Cu ions is close to the Fermi energy. Therefore, the coexistence of a 3D energy band with the 1D p_π band is expected [387].

In order to verify these suggestions dHvA measurements were carried out for single crystals of deuterated and undeuterated $(DMe-DCNQI)_2Cu$. Figure 4.43 shows the result for two orientations at $T = 0.05\,K$ [388]. The Fourier transformations of the data are shown in the insets. For the field applied along the c direction only one frequency of $\sim 477\,T$, denoted by α, is found. When the field is applied in the ab plane a more complicated dHvA signal with many different frequencies can be observed. For the direction of $\varphi = 26°$ shown in Fig. 4.43, where φ is the angle between the field and the [110] direction, the Fourier transform shows four peaks. They were ascribed to one low frequency (δ_2) with the corresponding second harmonic ($2\delta_2$) and to two higher ones (ϵ_5 and ϵ_6) with frequencies above 4000 T. The latter orbits are larger than the area of the first Brillouin zone and must, therefore, be connected with MB orbits.

The measured angular dependence of the different frequencies is quite complicated. Some frequencies could only be observed over a narrow angular range [388]. Selected frequencies and corresponding effective cyclotron masses at certain angles are given in Table 4.3. In spite of the variety of experimentally found orbits they can be understood fairly well with the calculated FS.

Table 4.3. Experimental and calculated values of dHvA frequencies and effective cyclotron masses in undeuterated $(DMe-DCNQI)_2Cu$. For the ϵ orbits no effective masses have been calculated. From [388]

label	direction	$F_{exp}(T)$	m_c^{exp}/m_e	$F_{cal}(T)$	m_b^{cal}/m_e
α	$\Theta = 0°$	477	3.5	477	4.0
β	$\Theta = 60°$	403	3.0	438	3.7
β	$\varphi = 0°$	489	3.9	467	3.6
γ	$\varphi = 0°$	353	3.4	404	3.9
δ_1	$\varphi = 45°$	432	3.9	533	4.3
δ_2	$\varphi = 25°$	1221	6.5	1340	6.2
ϵ_5	$\varphi = 25°$	5235	9.5	5310	–
ϵ_6	$\varphi = 25°$	4525	8.0	4180	–
ϵ_7	$\varphi = 33°$	7071	10	6450	–
ϵ_8	$\varphi = 33°$	7831	12	7610	–

The calculated tight-binding band structure based on the 20 K lattice parameters is shown in Fig. 4.44. Compared to other organic metals, unusually complicated dispersion relations are found which consist of 1D and 3D energy bands. Perpendicular to c two 1D FS sheets denoted by FS1 and FS2 exist. Within the approximation used the doubly degenerate FS1 is ideally

1D without any warping. The FS2, on the other hand, shows a considerable corrugation due to the overlap of the Cu-$3d$ orbitals. The 3D FS3 stems from the $3d$ Cu band. At the point P along the XU line (see Fig. 4.44) FS2 and FS3 touch each other.

A schematic projection of the 3D FS3 is shown in Fig. 4.45 [388]. The bold lines show the possible extremal orbits of this multiple-connected FS. For a field in the c direction only the orbit α is possible which has been seen only experimentally (Fig. 4.43a). For fields in the ab plane the calculation predicts a variety of dHvA frequencies some of which should only be visible over a narrow angular range. This is in principal agreement with the experimental observations. From the 3D FS3 alone one obtains absolute values of the dHvA frequencies α, β, γ, and δ_i in reasonable agreement with the experiment. Table 4.3 shows the measured and calculated frequencies in comparison.

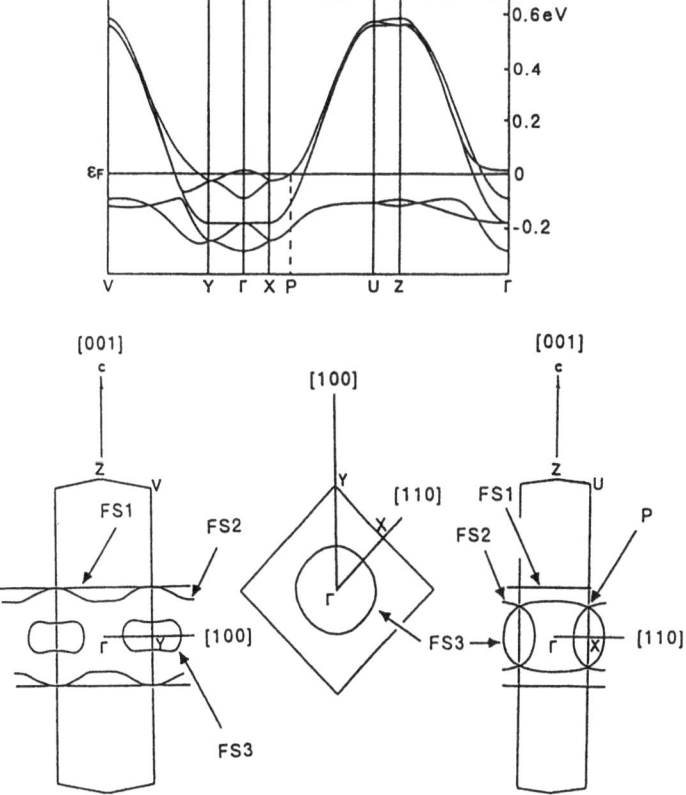

Fig. 4.44. Calculated band structure of (DMe–DCNQI)$_2$Cu. From [388]

The occurrence of the high oscillation frequencies ϵ_i can be understood by a possible MB orbit which extends from FS3 to FS2 over the point P where the two Fermi surfaces touch each other. Within this model the existence of the high frequencies can be explained. However, the absolute values given for selected orbits in Table 4.3 and the angular region where the orbits should be observable is not exactly in agreement with the experiment. This may be due to some oversimplifications in the band-structure calculations. Nevertheless, the gross experimentally observed features can be reproduced very well by the calculated FS. Recent first-principle LDA calculations [389] have reproduced the principal electronic structure obtained by the semiempirical tight-binding method shown in Fig. 4.44.

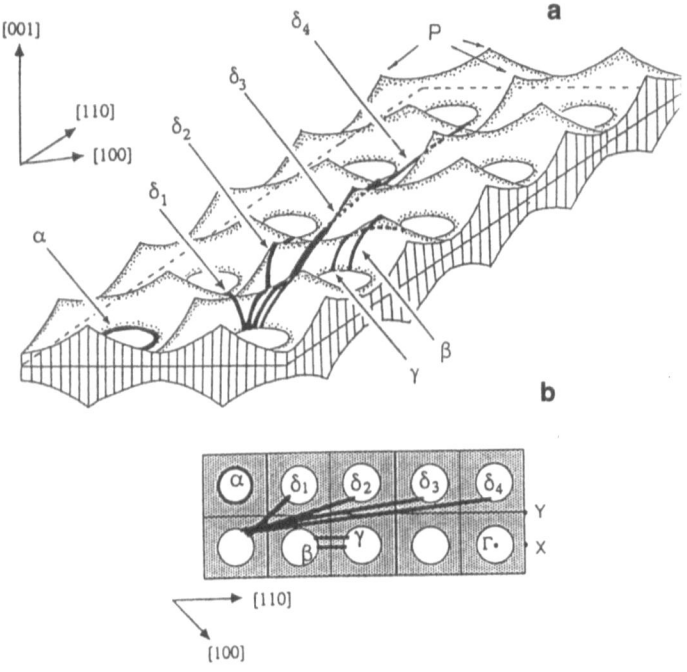

Fig. 4.45. (a) Schematic projection of the 3D FS of (DMe–DCNQI)$_2$Cu in the repeated zone scheme. (b) Cut of the FS through the *ab* plane. The proposed dHvA orbits are shown by the solid lines. From [388]

Finally, the extracted effective cyclotron masses, m_c^{exp}, given in Table 4.3 are also in surprisingly good agreement with the calculated band masses, m_b^{cal}. This might be fortuitous since an uncertainty of the absolute m_b^{cal} values of a factor of 2–3 is estimated [388]. However, the relatively large effective masses are understood by the substantial mixing between the organic p_π and the Cu 3d orbitals. No many-body effects have to be assumed to explain any unusually large experimental effective cyclotron masses.

Additional dHvA measurements of partially deuterated samples have revealed almost the same orbits and effective cyclotron masses as those mentioned for the undeuterated crystals [390]. The unchanged effective masses are somewhat surprising since in specific heat and susceptibility measurements an increased density of states for samples exhibiting the metal–insulator transition was found [391]. It was, therefore, assumed that strong electron correlations or electron–phonon interactions would enhance the effective mass of the conduction electrons [392]. The dHvA results, however, are clearly contradictory to these suggestions. Further studies are necessary to clarify this puzzle.

5. Conclusion

Since the first discovery of superconductivity in an organic material about 15 years ago many groups all over the world have made great efforts in synthesizing new materials with better conductivity and higher T_c. An important and necessary condition to reach the goal of tailoring compounds to the behavior required is a rigorous investigation and hopefully a deeper understanding of the physical properties of these fascinating synthetic metals. A large variety of experiments have been carried out. Thereby, an extraordinarily close collaboration between chemists and physicists was initiated.

Meanwhile more than 60 organic superconductors are now known and an enormous number of newly synthesized organic metals has been found. These materials revealed an unexpected large variety of different physical phenomena. Due to the low charge-carrier density the energy scale which determines the solid-state properties is comparatively small. Therefore, the ground state of the systems is extremely sensitive to external parameters such as pressure, thermal history, and magnetic fields. Consequently, the phase diagrams of organic metals include many different phases most of which occur within the parameter range available with standard laboratory equipment. This gives a unique opportunity to study bulk solid-state materials near extreme limits where quantum effects play an appreciable role.

The synthetic metals discussed here were all characterized by their electronic low dimensionality and, consequently, by their extreme anisotropies. Textbook like 1D metals are known to be unstable against a Peierls transition at low temperatures. Indeed, the 1D organic compounds based on TTF–TCNQ, TMTTF, and TMTSF comprise a whole variety of these ground states. Systematic studies during the last few years have enabled better understanding of these phases. A possible way to prevent the organic metals to become insulating due to the formation of a CDW or SDW is to increase the dimensionality. This can easily be accomplished by the application of pressure. Magnetic fields, on the other hand, reduce the effective dimension of the charge carrier path and can, therefore, induce SDW transitions in a metal. The experimental findings of these phenomena in the Bechgaard salts initiated the development of new theories which are now known as the standard model for field-induced spin density waves (FISDW).

The 1D organic metals are discussed currently as prime candidates for showing Luttinger liquid behavior. The missing of a sharp FS edge in photoemission studies especially suggested the non-Fermi-liquid picture. Continuous experimental efforts are being spent to verify the proposed physical characteristics of the theory.

The property governing the low-temperature thermodynamic and transport behavior of a metal is the electronic band structure and the topology of the FS. Therefore, the exact knowledge of the electronic structure determined both by band-structure calculations and verified by experimental studies is essential. Although the direct evaluation of 1D Fermi surfaces by dHvA or SdH measurements is principally not possible due to the absence of closed orbits, refined experiments measuring the angular dependence of the magnetoresistance seem to be a promising technique for FS investigations.

More direct FS information is available for 2D materials. Indeed, a large number of dHvA and SdH experiments have been done for these organic metals. In almost all cases a nearly perfect 2D FS cylinder was found. In the α, κ, and Θ phases additional 1D bands exist. These bands have been established by the observation of magnetic breakdown orbits. Therefore, the simple textbook-like FS allows a detailed check of existing and newly proposed magnetic-breakdown theories. For many quasi-2D organic metals the existence of AMROs proved a certain degree of three dimensionality. For some materials, especially the β-phase salts, this remaining inter-layer overlap could be exactly determined by dHvA measurements. The in-plane FS shape was often extractable from AMRO experiments. Consequently, for a variety of 2D salts the complete band structure has been mapped out experimentally. The agreement with relatively simple tight-binding band-structure calculations is surprisingly good.

The almost perfect 2D FS is responsible for the sometimes unusual behavior of the SdH or dHvA oscillations, like an enhanced anharmonicity or a seemingly field-dependent effective mass. A detailed theoretical understanding of the dHvA effect for 2D materials with a similar powerful equation such as the Lifshitz–Kosevich formula is still missing. The current investigations on magnetic quantum oscillations of 2D materials, especially comparative studies of magnetic (dHvA) and resistive (SdH) measurements, will hopefully help to clarify the open questions.

Some organic 2D metals, namely of the α-phase family, with an additional 1D band are unstable against a DW transition. This results in characteristic features in SdH and dHvA signals. Finally, AMRO experiments resolved the nesting vector connected with this transition. Usual methods were not able to resolve the nesting vector since the magnetic moment induced by the SDW transition is extremely small. The details of this ground state instability, however, are a current topic of controversial debate and still have to be clarified in further investigations.

To date the origin of the superconductivity in organic metals has been discussed controversially. The experimental situation is far from giving a consistent picture. Some experiments reveal unconventional superconducting properties while others reproduce the usual BCS-like behavior. dHvA measurements showed consistency with the conventional theory. Under certain assumptions the electron–phonon coupling constant could be extracted for some 2D materials and was in the correct range to account for the observed T_cs. On the other hand, cyclotron resonance studies seem to suggest an additional very strong electron–electron interaction. Whether these effects are important ingredients for the superconducting properties is currently debated.

The comparatively low critical temperatures of the known organic superconductors to date allow thorough investigations of the superconducting phase diagram of these bulk low-dimensional metals. Comprehensive experimental studies of the thermodynamic properties, the critical fields, and the mixed-state behavior covering almost the complete superconducting phase are possible. The results obtained are of fundamental importance with respect to the numerous theoretical predictions regarding the physical properties of low-dimensional strongly type-II superconductors.

Since the number of organic metals and superconductors are continuously increasing and due to the ever improving crystal quality, the near future certainly will give more important information on the electronic structure and superconducting properties of synthetic metals. The fermiology, the investigation of the FS topology, has become an important contributor to the understanding of these fascinating materials. Hopefully the future will see organic compounds with even higher superconducting transition temperatures. However, already the presently known materials have revealed an enormous richness of new and unusual physical properties which have stimulated many further experiments. Due to the relatively simple and now partially understood band structure, organic metals have become model substances for low-dimensional electronic systems.

References

[1] H. Inokuchi and H. Akamatsu in: *Solid State Physics*, Vol. 12, ed. F. Seitz and D. Turnbull (Academic Press, New York 1961), p. 93

[2] J. Ferraris, D. O. Cowan, V. Walatka, Jr., and J. H. Perlstein, J. Am. Chem. Soc. **95**, 948 (1973); L. B. Coleman, M. J. Cohen, D. J. Sandman, F. G. Yamagishi, A. F. Garito, and A. J. Heeger, Solid State Commun. **12**, 1125 (1973)

[3] R. E. Peierls, *Quantum Theory of Solids* (Oxford Univ. Press, London 1955)

[4] H. Fröhlich, Proc. Roy. Soc. London Ser. A **223**, 296 (1954)

[5] For reviews about 1D conductors see e. g.: H. J. Keller (ed.), *Low-Dimensional Cooperative Phenomena* (Plenum, New York 1974); S. Kagoshima, H. Nagasawa, and T. Sambongi, *One-Dimensional Conductors*, Springer Ser. Solid State Sci. Vol. 72 (Springer, Berlin, Heidelberg 1988)

[6] P. Delhaès in: *Lower-Dimensional Systems and Molecular Electronics*, ed. R. M. Metzger, P. Day, and G. C. Papavassiliou, NATO ASI Ser. B, Vol. 248 (Plenum, New York 1991), p. 43

[7] J. M. Luttinger, J. Math. Phys. **4**, 1154 (1963); E. H. Lieb and D. C. Mattis, J. Math. Phys. **6**, 304 (1965)

[8] B. Dardel, D. Malterre, M. Grioni, P. Weibel, Y. Baer, J. Voit, and D. Jérome, Europhys. Lett. **24**, 687 (1993)

[9] R. Liu, H. Ding, J. C. Campuzano, H. H. Wang, J. M. Williams, and K. D. Carlson, Phys. Rev. B **51**, 6155 (1995) & 13000 (1995)

[10] R. Itti, H. Mori, K. Ikeda, I. Hirabayashi, N. Koshizuka, and S. Tanaka, Physica C **185–189**, 2673 (1991); Mod. Phys. Lett. B **6**, 1785 (1992)

[11] W. A. Little, Phys. Rev. **A134**, 1416 (1964)

[12] D. Jérome, A. Mazaud, M. Ribault, and K. Bechgaard, J. Physique Lett. **41**, L95 (1980)

[13] For recent overviews of the discovered organic superconductors to date see: H. Mori, Int. J. Mod. Physics B **8**, 1 (1994); Proc. of the International Conference of Science and Technology of Synthetic Metals (ICSM) 1994, Seoul, Korea, Synth. Met. **69–70**, (1995)

[14] K. Bechgaard, K. Caneiro, M. Olsen, F. B. Rasmussen, and C. S. Jacobsen, Phys. Rev. Lett. **46**, 852 (1981)

[15] S. S. P. Parkin, E. M. Engler, R. R. Schumaker, R. Lagier, V. Y. Lee, J. C. Scott, and R. L. Greene, Phys. Rev. Lett. **50**, 270 (1983)

[16] J. M. Williams, A. M. Kini, H. H. Wang, K. D. Carlson, U. Geiser, L. K. Montgomery, G. J. Pyrka, D. M. Watkins, J. M. Kommers, S. J. Boryschuk, A. V. Strieby Crouch, W. K. Kwok, J. E. Schirber, D. L. Overmyer, D. Jung, and M.-H. Whangbo, Inorg. Chem. **29**, 3272 (1990)

[17] K. Kikuchi, M. Kikuchi, T. Namiki, K. Saito, I. Ikemoto, K. Murata, T. Ishiguro, and K. Kobayashi, Chem . Lett. **1987**, 931 (1987); K. Kikuchi,

Y. Honda, Y. Ishikawa, K. Saito, I. Ikemoto, K. Murata, H. Anzai, and T. Ishiguro, Solid State Commun. **66**, 405 (1988) and Refs therein

[18] G. C. Papavassiliou, G. A. Mousdis, J. S. Zambounis, A. Terzis, A. Hountas, B. Hilti, C. W. Mayer, and J. Pfeiffer, Synth. Metals **27**, B379 (1988)

[19] R. Kato, S. Aonuma, Y. Okano, H. Sawa, M. Tamura, M. Kinoshita, K. Oshima, A. Kobayashi, K. Bun, and H. Kobayashi, Synth. Metals **61**, 199 (1993); K. Oshima, H. Okuno, K. Kato, R. Murayama, R. Kato, A. Kobayashi, and H. Kobayashi, Synth. Metals **70**, 861 (1995)

[20] M. A. Beno, H. H. Wang, A. M. Kini, K. D. Carlson, U. Geiser, W. K. Kwok, J. E. Thompson, J. M. Williams, J. Ren, and M.-H. Whangbo, Inorg. Chem. **29**, 1599 (1990)

[21] S. Kahlich, D. Schweitzer, I. Heinen, S. E. Lan, B. Nuber, H. J. Keller, K. Winzer, and H. W. Helberg, Solid State Commun. **80**, 191 (1991)

[22] For a review see: P. Cassoux, L. Valade, H. Kobayashi, A. Kobayashi, R. A. Clark, and A. E. Underhill, Coord. Chem. Rev. **110**, 115 (1991)

[23] L. Brossard, M. Ribault, L. Valade, and P. Cassoux, Physica B **143**, 378 (1986); J. Phys. (Paris) **50**, 1521 (1989); A. Kobayashi, H. Kim, Y. Sasaki, R. Kato, H. Kobayashi, S. Moriyama, Y. Nishio, K. Kajita, and W. Sasaki, Chem. Lett. **1987**, 1819 (1987); A. Kobayashi, R. Kato, A. Miyamoto, T. Naito, H. Kobayashi, R. A. Clark, and A. E. Underhill, Chem. Lett. **1991**, 2163 (1991); H. Kobayashi, K. Bun, T. Naito, R. Kato, and A. Kobayashi, Chem. Lett. **1992**, 1909 (1992); H. Tajima, M. Inokuchi, A. Kobayashi, T. Ohta, R. Kato, H. Kobayashi, and H. Kuroda, Chem. Lett. **1993**, 1235 (1993)

[24] M. T. Béal-Monod, C. Bourbonnais, and V. J. Emery, Phys. Rev. B **34**, 7716 (1986); D. J. Scalapino, E. Loh, Jr., and J. E. Hirsch, Phys. Rev. B **35**, 6694 (1987)

[25] For recent articles and reviews see: R. L. Greene and P. M. Chaikin, Physica B **126**, 431 (1984); *Low-Dimensional Conductors and Superconductors*, ed. by D. Jérome and L. G. Caron, NATO ASI Ser. B, Vol. **155** (Plenum, New York 1987); D. Jérome, Solid State Commun. **92**, 89 (1994)

[26] P. W. Anderson, Science **256**, 1526 (1992) & Physica C **185–189**, 11 (1991); P. W. Anderson, G. Baskaran, Z. Zou, J. Wheatley, T. Hsu, B. S. Shastry, B. Doucot, S. Liang, Physica C **153–155**, 527 (1988)

[27] J. Sólyom, Adv. Phys. **28**, 201 (1979)

[28] T. Ishiguro and K. Yamaji, *Organic Superconductors* (Springer, Berlin, Heidelberg 1990)

[29] K. Oshima, T. Mori, H. Inokuchi, H. Urayama, H. Yamochi, and G. Saito, Phys. Rev. B **38**, 938 (1988)

[30] M. V. Kartsovnik, V. N. Laukhin, V. I. Nizhankovskii, and A. A. Ignat'ev, Pis'ma Zh. Eksp. Teor. Fiz. **47**, 302 (1988) [JETP Lett. **47**, 541 (1988)]; M. V. Kartsovnik, P. A. Kononovich, V. N. Laukhin, and I. F. Schegolev, Pis'ma Zh. Eksp. Teor. Fiz. **48**, 498 (1988) [JETP Lett. **48**, 541 (1988)]

[31] A recent review is given in: J. Wosnitza, Int. J. Mod. Phys. B **7**, 2707 (1993)

[32] P. J. van der Wel, J. Caulfield, J. Singleton, R. Corcoran, S. M. Hayden, M. Springford, W. Hayes, M. Kurmoo, and P. Day, Physica C **235–240**, 2453 (1994); P. J. van der Wel, J. Caulfield, S. M. Hayden, J. Singleton, M. Springford, P. Meeson, W. Hayes, M. Kurmoo, and P. Day, Synth. Metals **70**, 831 (1995)

[33] J. M. Williams, J. R. Ferraro, R. J. Thorn, K. D. Carlson, U. Geiser, H. H. Wang, A. M. Kini, and M.-H. Whangbo, *Organic Superconductors: Synthesis, Structure, Properties, and Theory* (Prentice Hall, Englewood Cliffs, 1992) and references therein

[34] R. P. Shibaeva, É. B. Yagubskii, E. E. Laukhina, and V. N. Laukhin, in: *The Physics and Chemistry of Organic Superconductors*, ed. G. Saito and S. Kagoshima (Springer, Berlin, Heidelberg, 1990), p. 342; K. D. Carlson, H. H. Wang, M. A.Beno, A. M. Kini, and J. M. Williams, Mol. Cryst. Liq. Cryst. **181**, 91 (1990)

[35] K. Bechgaard, C. S. Jacobsen, K. Mortensen, J. H. Pederson, and N. Thorup, Solid State Commun. **33**, 1119 (1980)

[36] L. Balicas, K. Behnia, W. Kang, E. Canadell, P. Auban-Senzier, D. Jérome, M. Ribault, and J. M. Fabre, J. Phys. I France **4**, 1539 (1994)

[37] See, for example: Proc. of the International Conference on Low Dimensional Conductors and Superconductors, Les Arcs, 1982 [J. Physique Colloq. **44**, C3 (1983)] and Proc. of the International Conference on the Physics and Chemistry of Low-Dimensional Synthetic Metals (ICSM '84), Abano Terme, 1984 [Mol. Cryst. Liq. Cryst. **119**, (1985)]

[38] N. Thorup, G. Rindorf, H. Soling, and K. Bechgaard, Acta Cryst. B **37**, 1236 (1981)

[39] P. Batail, L. Ouahab, J. B. Torrance, M. L. Pylman, and S. S. P. Parkin, Solid State Commun. **55**, 597 (1985)

[40] F. Wudl, E. Aharon-Shalom, D. Nalewajek, J. V. Waszczak, W. M. Walsh, Jr., L. W. Rupp, Jr., P. M. Chaikin, R. C. Lacoe, M. Burns, T. O. Poehler, J. M. Williams, and M. A. Beno, J. Chem. Phys. **76**, 5497 (1982); R. C. Lacoe, S. A. Wolf, P. M. Chaikin, F. Wudl, and E. Aharon-Shalom, Phys. Rev. B **27**, 1947 (1983)

[41] See, e. g.: R. Hoffmann, J. Chem. Phys. **39**, 1397 (1963); M.-H. Whangbo and R. Hoffmann, J. Am. Chem. Soc. **100**, 6093 (1978)

[42] C. S. Jacobsen, D. B. Tanner, and K. Bechgaard, Phys. Rev. B **28**, 7019 (1983)

[43] P. M. Grant, J. Physique **44**, C3–847 (1983)

[44] T. Ishiguro, T. Ukachi, K. Kato, K. Murata, M. Tokumoto, H. Tokumoto, H. Anzai, and G. Saito, J. Phys. Soc. Jpn. **52**, 1585 (1983); H. Schwenk, K. Andres, F. Wudl, and E. Aharon-Shalom, J. Physique **44**, C3–1041 (1983)

[45] For reviews about CDW phenomena see: G. Grüner, Rev. Mod. Phys. **60**, 1129 (1988); L. P. Gor'kov and G. Grüner (eds.), *Charge Density Waves in Solids* (North–Holland, Amsterdam 1989)

[46] A. W. Overhauser, Phys. Rev. Lett. **4**, 462 (1960)

[47] See, for example: D. Jérome and H. J. Schulz, Adv. Phys. **31**, 299 (1982)

[48] J. P. Pouget, S. K. Khanna, F. Denoyer, R. Comès, A. F. Garito, and A. J. Heeger, Phys. Rev. Lett. **37**, 437 (1976); K. S. Khanna, J. P. Pouget, R. Comès, A. F. Garito, and A. J. Heeger, Phys. Rev. B **16**, 1468 (1977)

[49] S. Kagoshima, T. Ishiguro, and H. Anzai, J. Phys. Soc. Jpn. **41**, 2061 (1976)

[50] J. Kondo and K. Yamaji, J. Phys. Soc. Jpn. **43**, 424 (1987)

[51] G. Soda, D. Jérome, M. Weger, J. Alizou, J. Gallice, H. Robert, J. M. Fabre, and L. Giral, J. Physique **38**, 931 (1977)

[52] J. B. Torrance, Y. Tomkiewicz, and B. D. Silverman, Phys. Rev. B **15**, 4738 (1977)

[53] J. C. Scott, H. J. Pedersen, and K. Bechgaard, Phys. Rev. Lett. **45**, 2125 (1980)

[54] K. Mortensen, Y. Tomkiewicz, T. D. Schultz, and E. M. Engler, Phys. Rev. Lett. **46**, 1234 (1981)

[55] K. Mortensen, Y. Tomkiewicz, and K. Bechgaard, Phys. Rev. B **25**, 3319 (1982)

[56] H. J. Pedersen, J. C. Scott, and K. Bechgaard, Solid State Commun. **35**, 207 (1980)

[57] W. M. Walsh, Jr., F. Wudl, G. A. Thomas, D. Nalewajek, J. J. Hauser, P. A. Lee, T. Poehler, Phys. Rev. Lett. **45**, 829 (1980)

[58] A. Andrieux, D. Jérome, and K. Bechgaard, J. Physique Lett. **42**, L87 (1981); J. C. Scott, H. J. Pedersen, and K. Bechgaard, Phys. Rev. B **24**, 475 (1981)

[59] J. B. Torrance, H. J. Pedersen, and K. Bechgaard, Phys. Rev. Lett. **49**, 881 (1982)

[60] J. M. Delrieu, M. Roger, Z. Toffano, E. Wope Mbougue, R. Saint James, and K. Bechgaard, Physica B **143**, 412 (1986)

[61] T. Takahashi, Y. Maniwa, H. Kawamura, and G. Saito, Physica B **143**, 417 (1986)

[62] D. Jérome, Science **252**, 1509 (1991)

[63] V. J. Emery, R. Bruinsma, and S. Barišić, Phys. Rev. Lett. **48**, 1039 (1982)

[64] J. P. Pouget, R. Moret, R. Comès, K. Bechgaard, J. M. Fabre, and L. Giral, Mol. Cryst. Liq. Cryst. **79**, 129 (1982)

[65] A. Maaroufi, S. Flandrois, G. Fillion, and J. P. Morand, Mol. Cryst. Liq. Cryst. **119**, 311 (1985); A. Maaroufi, S. Flandrois, C. Coulon, P. Delhaes, J. P. Morand, and G. Fillion, J. Physique Colloq. C3, **44**, 1091 (1983)

[66] S. S. P. Parkin, J. C. Scott, J. B. Torrance, and E. M. Engler, J. Physique Colloq. C3, **44**, 1111 (1983)

[67] F. Creuzet, D. Jérome, and A. Moradpour, Mol. Cryst. Liq. Cryst. **119**, 297 (1985)

[68] T. Takahashi, F. Creuzet, D. Jérome, and J. M. Fabre, J. Physique Colloq. C3, **44**, 1095 (1983)

[69] S. S. P. Parkin, F. Creuzet, M. Ribault, D. Jérome, K. Bechgaard, and J. M. Fabre, Mol. Cryst. Liq. Cryst. **79**, 249 (1982); F. Creuzet, S. S. P. Parkin, D. Jérome, K. Bechgaard, and J. M. Fabre, J. Physique Colloq. C3, **44**, 1099 (1983); S. S. P. Parkin, F. Creuzet, D. Jérome, J. M. Fabre, and K. Bechgaard, J. Physique **44**, 975 (1983)

[70] J. P. Pouget, G. Shirane, K. Bechgaard, and J. M. Fabre, Phys. Rev. B **27**, 5203 (1983)

[71] J. P. Pouget, R. Moret, R. Comes, and K. Bechgaard, J. Physique Lett. **42**, L543 (1981)

[72] J. F. Kwak, J. E. Schirber, R. L. Greene, and E. M. Engler, Phys. Rev. Lett. **46**, 1296 (1981); J. F. Kwak, J. E. Schirber, R. L. Greene, and E. M. Engler, Mol. Cryst. Liq. Cryst. **79**, 111 (1982)

[73] M. Ribault, D. Jérome, J. Tuchendler, C. Weyl, and K. Bechgaard, J. Physique Lett. **44**, L953 (1983); P. M. Chaikin, M.-Y. Choi, J. F. Kwak, J. S. Brooks, K. P. Martin, M. J. Naughton, E. M. Engler, and R. L. Greene, Phys. Rev. Lett. **51**, 2333 (1983)

[74] K. v. Klitzing, G. Dorda, and M. Pepper, Phys. Rev. B **28**, 4886 (1983)

[75] L. P. Gor'kov and A. G. Lebed', J. Physique Lett. **45**, L433 (1984)

[76] G. Montambaux, M. Héritier, and P. Lederer, J. Physique Lett. **45**, L533 (1984); M. Héritier, G. Montambaux, and P. Lederer, J. Physique Lett. **45**, L943 (1984); P. M. Chaikin, Phys. Rev. B **31**, 4770 (1985); K. Yamaji, J. Phys. Soc. Jpn. **54**, 1034 (1985); M. Ya. Azbel, P. Bak, and P. M. Chaikin, Phys. Lett. A **117**, 92 (1986); K. Maki, Phys. Rev. B **33**, 4826 (1986)

[77] J. R. Cooper, W. Kang, P. Auban, G. Montambaux, D. Jérome, and K. Bechgaard, Phys. Rev. Lett. **63**, 1984 (1989); S. T. Hannahs, J. S. Brooks, W. Kang, L. Y. Chiang, and P. M. Chaikin, Phys. Rev. Lett. **63**, 1988 (1989)

[78] D. Poilblanc, G. Montambaux, M. Héritier, and P. Lederer, Phys. Rev. Lett. **58**, 270 (1987)

[79] P. M. Chaikin, J. S. Brooks, S. T. Hannahs, W. Kang, G. Montambaux, and L. Y. Chiang in: *The Physics and Chemistry of Organic Superconductors*, ed. G. Saito and S. Kagoshima (Springer, Berlin, Heidelberg 1990), p. 81

[80] R. V. Chamberlin, M. J. Naughton, X. Yan, L. Y. Chiang, S.-Y. Hsu, and P. Chaikin, Phys. Rev. Lett. **60**, 1189 (1988); M. J. Naughton, R. V. Chamberlin, X. Yan, S.-Y. Hsu, L. Y. Chiang, M. Ya. Azbel, and P. M. Chaikin, Phys. Rev. Lett. **61**, 621 (1988); P. M. Chaikin, M. Ya. Azbel, M. J. Naughton, R. V. Chamberlin, X. Yan, S. Hsu, and L. Y. Chiang, Synth. Metals **27**, B163 (1988)

[81] R. C. Yu, L. Chiang, R. Upasani, and P. M. Chaikin, Phys. Rev. Lett. **65**, 2458 (1990)

[82] K. Machida, Y. Hori, and M. Nakano, Phys. Rev. Lett. **70**, 61 (1993); K. Machida, Y. Hasegawa, M. Kohmoto, V. M. Yakovenko, Y. Hori, and K. Kishigi, Phys. Rev. B **50**, 921 (1994)

[83] A. G. Lebed' and P. Bak, Phys. Rev. B **40**, 11433 (1989); L. P. Gor'kov and A. G. Lebed', Phys. Rev. B **51**, 1362 (1995).

[84] T. Osada, S. Kagoshima, and N. Miura, Phys. Rev. Lett. **69**, 1117 (1992)

[85] R. H. McKenzie, Phys. Rev. Lett. **74**, 5140 (1995)

[86] V. M. Yakovenko, Zh. Eksp. Teor. Fiz. **93**, 627 (1987) [Sov. Phys. JETP **66**, 355 (1987)]

[87] W. Kang, S. T. Hannahs, L. Y. Chiang, R. Upasani, and P. M. Chaikin, Phys. Rev. Lett. **65**, 2812 (1990)

[88] M. Basletić, N. Biškup, B. Korin-Hamzić, S. Tomić, A. Hamzić, K. Bechgaard, and J. M. Fabre, Europhys. Lett. **22**, 279 (1993)

[89] W. Kang, S. T. Hannahs, and P. M. Chaikin, Phys. Rev. Lett. **70**, 3091 (1993)

[90] S. K. McKernan, S. T. Hannahs, U. M. Scheven, G. M. Danner, and P. M. Chaikin, Phys. Rev. Lett. **75**, 1630 (1995)

[91] P. M. Chaikin, M.-Y. Choi, J. F. Kwak, J. S. Brooks, K. P. Martin, M, J. Naughton, E. M. Engler, and R. L. Greene, Phys. Rev. Lett. **51**, 2333 (1983); H. Schwenk, S. S. P. Parkin, R. Schumaker, R. L. Greene, and D. Schweitzer, Phys. Rev. Lett. **56**, 667 (1986); J. P. Ulmet, P. Auban, A. Khmou, S. Askenazy, and A. Moradpour, J. Physique Lett. **46**, L–535 (1985); J. P. Ulmet, A. Khmou, and L. Bachere, Physica B **143**, 400 (1986); A. Audouard, J. P. Ulmet, and J. M. Fabre, Synth. Metals **70**, 751 (1995)

[92] A. Audouard, F. Goze, S. Dubois, J. P. Ulmet, L. Brossard, S. Askenazy, S. Tomić, and J. M. Fabre, Europhys. Lett. **25**, 363 (1994); A. Audouard, F. Goze, S. Dubois, J. P. Ulmet, L. Brossard, S. Askenazy, and J. M. Fabre, Phys. Rev. B **50**, 12726 (1994); A. Audouard, F. Goze, J. P. Ulmet, L. Brossard, S. Askenazy, and J. M. Fabre, Synth. Metals **70**, 739 (1995); Physica B **211**, 303 (1995)

[93] X. Yan, M. J. Naughton, R. V. Chamberlin, S. Y. Hsu, L. Y. Chiang, J. S. Brooks, and P. M. Chaikin, Phys. Rev. B **36**, 1799 (1987); X. Yan, M. J. Naughton, R. V. Chamberlin, L. Y. Chiang, S. Y. Hsu, and P. M. Chaikin, Synth. Metals **27**, B145 (1988)

[94] T. Osada, N. Miura, and G. Saito, Solid State Commun. **60**, 441 (1986); Physica B **143**, 403 (1986)

[95] X. D. Shi, W. Kang, and P. M. Chaikin, Phys. Rev. B **50**, 1984 (1994)

[96] A. G. Lebed, Phys. Rev. Lett. **74**, 4903 (1995)

[97] K. Yamaji, J. Phys. Soc. Jpn. **56**, 1101 (1987)

[98] S. S. P. Parkin, M. Ribault, D. Jérome, and K. Bechgaard, J. Phys. C **14**, 5305 (1981)

[99] T. Takahashi, D. Jérome, and K. Bechgaard, J. Physique Lett. **43**, L565 (1982)

[100] K. Andres, F. Wudl, D. B. McWhan, G. A. Thomas, D. Nalewajek, and A. L. Stevens, Phys. Rev. Lett. **45**, 1449 (1980)

[101] R. L. Greene and E. M. Engler, Phys. Rev. Lett. **45**, 1587 (1980)

[102] See e. g.: W. Buckel, *Supraleitung*, 5th. edition, (VCH, Weinheim 1994)

[103] K. Murata, M. Tokumoto, H. Anzai, K. Kajimura, and T. Ishiguro, Jpn. J. Appl. Phys. **26**, 1367 (1987)

[104] D. R. Tilley, Proc. Phys. Soc. London **86**, 289 (1965); E. I. Kats, Sov. Phys. JETP **29**, 897 (1969)

[105] R. C. Morris, R. V. Coleman, and R. Bhandari, Phys. Rev B **5**, 895 (1972)

[106] R. A. Klemm and J. R. Clem, Phys. Rev. B **21**, 1868 (1980)

[107] W. E. Lawrence and S. Doniach, in *Proc. of the 12th Intern. Conf. on Low Temp. Physics, Kyoto 1970* ed. by E. Kanda (Academic Press of Japan, Tokyo 1971), Vol. 12, p. 361

[108] A. M. Clogston, Phys. Rev. Lett. **9**, 266 (1962)

[109] L. P. Gor'kov and D. Jérome, J. Physique Lett. **46**, L643 (1985)

[110] R. Brusetti, M. Ribault, D. Jérome, and K. Bechgaard, J. Physique **43**, 801 (1982)

[111] D. Mailly, M. Ribault, and K. Bechgaard, J. Physique **44**, C3–1037 (1983)

[112] K. Murata, H. Anzai, G. Saito, K. Kajimura, and T. Ishiguro, J. Phys. Soc. Jpn. **50**, 3529 (1981)

[113] P. Garoche, R. Brusetti, D. Jérome, and K. Bechgaard, J. Physique Lett. **43**, L147 (1982)

[114] C. More, G. Roger, J. P. Sorbier, D. Jérome, M. Ribault, and K. Bechgaard, J. Physique Lett. **42**, L313 (1981)

[115] A. Fournel, C. More, G. Roger, J. P. Sorbier, and C. Blanc, J. Physique **44**, C3–879 (1983)

[116] H. Bando, K. Kajimura, H. Anzai, T. Ishiguro, and G. Saito, Mol. Cryst. Liq. Cryst. **119**, 41 (1985)

[117] M. Takigawa, H. Yasuoka, and G. Saito, J. Phys. Soc. Jpn. **56**, 873 (1987)

[118] L. C. Hebel and C. P. Slichter, Phys. Rev. **113**, 1504 (1959)

[119] For recent experiments on UPt$_3$ see: H. v. Löhneysen, Physica B **197**, 551 (1994); The theoretical aspects are discussed in: J. A. Sauls, Adv. Phys. **43**, 113 (1994)

[120] Y. Hasegawa and H. Fukuyama, J. Phys. Soc. Jpn. **55**, 3978 (1986); J. Phys. Soc. Jpn. **56**, 877 (1987)

[121] J. E. Eldridge, C. C. Homes, F. E. Bates, and G. S. Bates, Phys. Rev. B **32**, 5156 (1985)

[122] D. Djurek, D. Jérome, and K. Bechgaard, J. Phys. C **17**, 4179 (1984)

[123] M.-Y. Choi, P. M. Chaikin, and R. L. Greene, Phys. Rev. B **34**, 7727 (1986)

[124] S. Tomić, D. Jérome, D. Mailly, M. Ribault, and K. Bechgaard, J. Physique **44**, C3–1075 (1983); C. Coulon, P. Delhaès, J. Amiell, J. P. Manceau, J. M. Fabre, and L. Giral, J. Physique **43**, 1721 (1982)

[125] P. C. W. Leung, T. J. Emge, M. A. Beno, H. H. Wang, J. M. Williams, V. Patrick, and P. Coppens, J. Am. Chem. Soc. **107**, 6184 (1985)

[126] J. M. Williams, H. H. Wang, T. J. Emge, U. Geiser, M. A. Beno, P. C. W. Leung, K. D. Carlson, R. J. Thorn, A. J. Schultz, and M.-H. Whangbo in: *Prog. Inorg. Chem.*, S. J. Lippard (ed.) (John Wiley, New York 1987) Vol. 35, p. 51

[127] R. P. Shibaeva, V. F. Kaminskii, and V. K. Bel'skii, Sov. Phys. Crystallogr. **29**, 638 (1984) [Kristallografiya **29**, 1089 (1984)]; J. M. Williams, T. J. Emge, H. H. Wang, M. A. Beno, P. T. Copps, L. N. Hall, K. D. Carlson, and G. W. Crabtree, Inorg. Chem. **23**, 2558 (1984)

[128] H. Kobayashi, R. Kato, A. Kobayashi, G. Saito, M. Tokumoto, H. Anzai, and
      T. Ishiguro, Chem Lett. **1985**, 1293 (1985)
[129] É. B. Yagubskii, I. F. Schegolev, V. N. Laukhin, P. A. Kononovich, M. V.
      Kartsovnik, A. V. Zvarykina, and L. I. Buravov, JETP Lett. **39**, 12 (1984)
      [Pis'ma Zh. Eksp. Teor. Fiz. **39**, 12 (1984)]
[130] J. M. Williams, H. H. Wang, M. A. Beno, T. J. Emge, L. M. Sowa, P. T.
      Copps, F. Behroozi, L. N. Hall, K. D. Carlson, and G. W. Crabtree, Inorg.
      Chem. **23**, 3839 (1984)
[131] T. J. Emge, H. H. Wang, M. A. Beno, P. C. W. Leung, M. A. Firestone, H.
      C. Jenkins, J. D. Cook, K. D. Carlson, J. M. Williams, E. L. Venturini, L. J.
      Azevedo, and J. E. Schirber, Inorg. Chem. **24**, 1736 (1985)
[132] T. J. Emge, P. C. W. Leung, M. A. Beno, A. J. Schultz, H. H. Wang, L. M.
      Sowa, J. M. Williams, Phys. Rev. B. **30**, 6780 (1984)
[133] Y. Nogami, S. Kagoshima, T. Sugano, and G. Saito, Synth. Metals **16**, 367
      (1986)
[134] F. Creuzet, G. Creuzet, D. Jérome, D. Schweitzer, and H. J. Keller, J.
      Physique Lett. **46**, L1079 (1985)
[135] H. Veith, C.-P. Heidmann, F. Gross, A. Lerf, K. Andres, and D. Schweitzer,
      Solid State Commun. **56**, 1015 (1985); K. Murata, M. Tokumoto, H. Anzai,
      H. Bando, G. Saito, K. Kajimura, and T. Ishiguro, J. Phys. Soc. Jpn. **54**,
      1236 & 2084 (1985); V. N. Laukhin, E. É. Kostyuchenko, Y. V. Sushko, I. F.
      Schegolev, and É. B. Yagubskii, Pis'ma Zh. Eksp. Teor. Fiz. **41**, 68 (1985)
      [JETP Lett. **41**, 81 (1985)]
[136] V. B. Ginodman, A. V. Gudenko, P. A. Kononovich, V. N. Laukhin, and I.
      F. Schegolev, Pis'ma Zh. Eksp. Teor. Fiz. **44**, 523 (1986) [JETP Lett. **44**, 673
      (1986)]; I. F. Schegolev, Jpn. J. Appl. Phys. **26**, Suppl. 26–3, 1972 (1987); W.
      Kang, D. Jérome, C. Lenoir, and P. Batail, Synth. Metals **27**, A353 (1988)
[137] H. Urayama, H. Yamochi, G. Saito, K. Nozawa, T. Sugano, M. Kinoshita,
      S. Saito, K. Oshima, A. Kawamoto, and J. Tanaka, Chem. Lett. **1988**, 55
      (1988)
[138] H. Urayama, H. Yamochi, G. Saito, S. Sato, A. Kawamoto, J. Tanaka, T.
      Mori, Y. Maruyama, and H. Inokuchi, Chem. Lett. **1988**, 463 (1988)
[139] A. Kobayashi, R. Kato, H. Kobayashi, S. Moriyama, Y. Nishino, K. Kajita,
      and W. Sasaki, Chem. Lett. **1987**, 459 (1987)
[140] M. Oshima, H. Mori, G. Saito, and K. Oshima, Chem. Lett. **1989**, 1159 (1989)
[141] H. Mori, S. Tanaka, M. Oshima, G. Saito, T. Mori, Y. Murayama, and H.
      Inokuchi, Bull. Chem. Soc. Jpn. **63**, 2183 (1990); H. Mori, S. Tanaka, K.
      Oshima, G. Saito, T. Mori, Y. Maruyama, and H. Inokuchi, Synth. Metals
      41–43, 2013 (1991)
[142] N. D. Kushch, L. I. Buravov, M. V. Kartsovnik, V. N. Laukhin, S. I. Pesotskii,
      R. P. Shibaeva, L. P. Rozenberg, É. B. Yagubskii, and A. V. Zvarikina, Synth.
      Met. **46**, 271 (1992)
[143] L. I. Buravov, N. D. Kushch, V. N. Laukhin, A. G. Khomenko, É. B. Yagub-
      skii, M. V. Kartsovnik, A. E. Kovalev, L. P. Rozenberg, R. P. Shibaeva, M.
      A. Tanatar, V. S. Yefanov, V. V. Dyakin, and V. A. Bondarenko, J. Phys. I
      France **4**, 441 (1994)
[144] T. Sasaki, H. Ozawa, H. Mori, S. Tanaka, T. Fukase, and N. Toyota, J. Phys.
      Soc. Jpn. **65**, to be published
[145] K. Bender, I. Hennig, D. Schweitzer, K. Dietz, H. Endres, H. J. Keller, Mol.
      Cryst. Liq. Cryst. **108**, 359 (1984)

[146] H. H. Wang, K. D. Carlson, U. Geiser, W. K. Kwok, M. D. Vashon, J. E. Thompson, N. F. Larsen, G. D. McCabe, R. S. Hulscher, and J. M. Williams, Physica C **166**, 57 (1990)

[147] H. Kobayashi, R. Kato, A. Kobayashi, Y. Nishio, K. Kajita, and W. Sasaki, Chem. Lett. **1986**, 789 & 833 (1986); K. Kajita, Y. Nishio, S. Moriyama, W. Sasaki, R. Kato, H. Kobayashi, and A. Kobayashi, Solid State. Commun. **64**, 1279 (1987)

[148] R. Kato, H. Kobayashi, A. Kobayashi, Y. Nishio, K. Kajita, and W. Sasaki, Chem. Lett. **1986**, 957 (1986)

[149] A. Kobayashi, R. Kato, H. Kobayashi, S. Moriyama, Y. Nishio, K. Kajita, and W. Sasaki, Chem. Lett. **1986**, 2017 (1986); H. Kobayashi, R. Kato, A. Kobayashi, T. Mori, H. Inokuchi, Y. Nishio, K. Kajita, and W. Sasaki, Synth. Metals **27**, A289 (1988)

[150] A. I. Schegolev, V. N. Laukhin, A. G. Khomenko, M. V. Kartsovnik, R. P. Shibaeva, L. P. Rozenberg, and A. E. Kovalev, J. Phys. I France **2**, 2123 (1992)

[151] H. H. Wang, M. A. Beno, K. D. Carlson, U. Geiser, A. M. Kini, L. K. Montgomery, J. E. Thompson, and J. M. Williams, in: *Organic Superconductivity*, ed. V. Z. Kresin and W. A. Little (Plenum, New York, 1990), p. 51

[152] T. Mori, A. Kobayashi, Y. Sasaki, H. Kobayashi, G. Saito, and H. Inokuchi, Bull. Chem. Soc. Jpn. **57**, 627 (1984)

[153] T. Mori, A. Kobayashi, Y. Sasaki, H. Kobayashi, G. Saito, and H. Inokuchi, Chem. Lett. **1984**, 957 (1984)

[154] M.-H. Whangbo, J. M. Williams, P. C. W. Leung, M. A. Beno, T. J. Emge, H. H. Wang, K. D. Carlson, and G. W. Crabtree, J. Am. Chem. Soc. **107**, 5815 (1985)

[155] M.-H. Whangbo, J. J. Novoa, D. Jung, J. M. Williams, A. M. Kini, H. H. Wang, U. Geiser, M. A. Beno, and K. D. Carlson, in: *Organic Superconductivity*, ed. V. Z. Kresin and W. A. Little (Plenum, New York, 1990), p. 243

[156] E. Canadell, private communication

[157] T. J. Emge, H. H. Wang, U. Geiser, M. A. Beno, K. S. Webb, and J. M. Williams, J. Am. Chem. Soc. **108**, 3849 (1986)

[158] A. J. Schultz, H. H. Wang, J. M. Williams, and A. J. Filhol, J. Am. Chem. Soc. **108**, 7853 (1986); A. J. Schultz, M. A. Beno, H. H. Wang, and J. M. Williams, Phys. Rev. B **33**, 7823 (1986)

[159] J. Kübler, M. Weger, and C. B. Sommers, Solid State Commun. **62**, 801 (1987)

[160] J. Kübler and C. B. Sommers, in: *The Physics and Chemistry of Organic Superconductors*, ed. G. Saito and S. Kagoshima (Springer, Berlin, Heidelberg, 1990), p. 208; M. Weger, A. Novack, D. Schweitzer, J. Kübler, J. M. van Bentum, and C. B. Sommers, in: *Organic Superconductivity*, ed. V. Z. Kresin and W. A. Little (Plenum, New York, 1990), p. 211

[161] K. Oshima, H. Urayama, H. Yamochi, and G. Saito, Physica C **153–155**, 1148 (1988)

[162] H. Urayama, H. Yamochi, G. Saito, S. Sato, T. Sugano, M. Kinoshita, A. Kawamato, J. Tanaka, T. Inabe, T. Mori, Y. Maruyama, H. Inokuchi, and K. Oshima, Synth. Metals **27**, A393 (1988)

[163] A. Kobayashi, R. Kato, H. Kobayashi, S. Moriyama, S. Nishio, K. Kajita, and W. Sasaki, Chem Lett. **1987**, 459 (1987)

[164] N. W. Ashcroft and N. D. Mermin, *Solid State Physics* (Saunders College, Philadelphia 1976). For the definition of hole and electron orbits see Chap. 12

[165] H. Kuroda, K. Yakushi, H. Tajima, A. Ugawa, Y. Okawa, A. Kobayashi, R. Kato, H. Kobayashi, and G. Saito, Synth. Metals **27**, A491 (1988); M. Tamura, H. Tajima, K. Yakushi, H. Kuroda, A. Kobayashi, R. Kato, and H. Kobayashi, J. Phys. Soc. Jpn. **60**, 3861 (1991)

[166] B. Koch, H. P. Geserich, W. Ruppel, D. Schweitzer, K. H. Dietz, and H. J. Keller, Phys. Rev. B **31**, 3138 (1985); C. S. Jacobsen, J. M. Williams, and H. H. Wang, Solid State Commun. **54**, 937 (1985)

[167] T. Sasaki, N. Toyota, M. Tokumoto, N. Kinoshita, and H. Anzai, Solid State Commun. **75**, 93 (1990); N. Kinoshita, M. Tokumoto, and H. Anzai, J. Phys. Soc. Jpn. **60**, 2131 (1991)

[168] F. L. Pratt, T. Sasaki, N. Toyota, and K. Nagamine, Phys. Rev. Lett. **74**, 3892 (1995)

[169] L. Ducasse and A. Fritsch, Solid State Commun. **91**, 201 (1994)

[170] K. Yamaji, J. Phys. Soc. Jpn. **58**, 1520 (1989); K. Yamaji, in *The Physics and Chemistry of Organic Superconductors*, ed. G. Saito and S. Kagoshima (Springer, Berlin, Heidelberg 1990), p. 216

[171] R. Yagi, Y. Iye, T. Osada, and S. Kagoshima, J. Phys. Soc. Jpn. **59**, 3069 (1990); Y. Kurihara, J. Phys. Soc. Jpn. **61**, 975 (1992)

[172] M. V. Kartsovnik, V. N. Laukhin, S. I. Pesotskii, I. F. Schegolev, and V. M. Yakovenko, J. Phys. I France **2**, 89 (1992)

[173] D. E. Farrell, C. J. Allen, R. C. Haddon, and S. V. Chichester, Phys. Rev. B **42**, 8694 (1990)

[174] J. C. Martinez, S. H. Brongersma, A. Koshelev, B. Ivlev, P. H. Kes, R. P. Griessen, D. G. de Groot, Z. Tarnavski, and A. A. Menovski, Phys. Rev. Lett. **69**, 2276 (1992)

[175] J. E. Schirber, L. J. Azevedo, J. F. Kwak, E. L. Venturini, P. C. W. Leung, M. A. Beno, H. H. Wang, and J. M. Williams, Phys. Rev. B **33**, 1987 (1986)

[176] H. Tanino, K. Kato, M. Tokumoto, H. Anzai, and G. Saito, J. Phys. Soc. Jpn. **54**, 2390 (1985)

[177] J. E. Schirber, E. L. Venturini, A. M. Kini, H. H. Wang, J. R. Whitworth, and J. M. Williams, Physica C **152**, 157 (1988)

[178] M. Lang, R. Modler, F. Steglich, N. Toyota, and T. Sasaki, Physica B **194–196**, 2005 (1994)

[179] M. Kund, H. Veith, H. Müller, K. Andres, and G. Saito, Physica C **221**, 119 (1994)

[180] J. E. Schirber, D. L. Overmyer, J. M. Williams, A. M. Kini, and H. H. Wang, Physica C **170**, 231 (1990)

[181] H. Schwenk, C.-P. Heidmann, F. Gross, E. Hess, K. Andres, D. Schweitzer, and H. J. Keller, Phys. Rev. B **31**, 3138 (1985)

[182] M. Tokumoto, H. Anzai, K. Takahashi, K. Murata, N. Kinoshita, and T. Ishiguro, Synth. Metals **27**, A305 (1988)

[183] S. Wanka, D. Beckmann, J. Wosnitza, E. Balthes, D. Schweitzer, W. Strunz, and H. J. Keller, Phys. Rev. B, to be published

[184] D. A. Brawner, A. Schilling, H. R. Ott, R. J. Haug, K. Ploog, and K. von Klitzing, Phys. Rev. Lett. **71**, 785 (1993)

[185] G. Blatter, B. Ivlev, and H. Nordberg, Phys. Rev. B **48**, 10448 (1993)

[186] W. J. Skocpol and M. Tinkham, Rep. Prog. Phys. **38**, 1049 (1975)

[187] U. Welp, S. Fleshler, W. K. Kwok, R. A. Klemm, V. M. Vinokur, J. Downey, B. Veal, and G. W. Crabtree, Phys. Rev. Lett. **67**, 3180 (1991)

[188] S. Ullah and A. T. Dorsey, Phys. Rev. Lett. **65**, 2066 (1990); Phys. Rev. B **44**, 262 (1991); R. Ikeda, T. Ohmi, and T. Tsuneto, J. Phys. Soc. Jpn. **58** 1377 (1989); **59** 1397 (1991)

[189] M. Lang, F. Steglich, N. Toyota, and T. Sasaki, Phys. Rev. B **49**, 15227 (1994)

[190] M. Tokumoto, H. Bando, H. Anzai, G. Saito, K. Murata, K. Kajimura, and T. Ishiguro, J. Phys. Soc. Jpn. **54**, 869 (1985)

[191] C.-P. Heidmann, K. Andres, and D. Schweitzer, Physica B **143**, 357 (1986)

[192] J. Wosnitza, N. Herrmann, X. Liu, and D. Schweitzer, Synth. Metals **70**, 829 (1995)

[193] M. Tinkham, Phys. Rev. **129**, 2413 (1963)

[194] M. Tinkham, *Introduction to Superconductivity* (McGraw–Hill, New York 1975)

[195] K. D. Carlson, G. W. Crabtree, L. Nuñez, H. H. Wang, M. A. Beno, U. Geiser, M. A. Firestone, K. S. Webb, and J. M. Williams, Solid State Commun. **57**, 89 (1986); D. Schweitzer, P. Bele, H. Brunner, E. Gogu, U. Haeberlen, I. Hennig, I. Klutz, R. Świetlik, and H. J. Keller, Z. Phys. B **67**, 489 (1987); D. Schweitzer, K. Polychroniadis, T. Klutz, H. J. Keller, I. Hennig, I. Heinen, U. Haeberlein, E. Gogu, and S. Gärtner, Synth. Metals **27**, A465 (1988); T. Sasaki, N. Toyota, M. Hasumi, T. Osada, S. Kagoshima, H. Anzai, M. Tokumoto, and N. Kinoshita, J. Phys. Soc. Jpn. **58**, 3477 (1989); J. S. Brooks, S. Uji, H. Aoki, T. Terashima, M. Tokumoto, N. Kinoshita, Y. Tanaka, and H. Anzai, Synth. Metals **70**, 839 (1995)

[196] Y. Yeshurun and A. P. Malozemoff, Phys. Rev. Lett. **60**, 2202 (1988); A. P. Malozemoff, T. K. Worthington, Y. Yeshurun, F. Holtzberg, and P. H. Kes, Phys. Rev. B **38**, 7203 (1988)

[197] P. W. Anderson, Phys. Rev. Lett. **9**, 309 (1962); P. W. Anderson and Y. B. Kim, Rev. Mod. Phys. **36**, 39 (1964); A. M. Campbell and J. E. Evetts, Adv. Phys. **21**, 199 (1972).

[198] J. Wosnitza, X. Liu, D. Schweitzer, and H. J. Keller, Phys. Rev. B **50**, 12747 (1994)

[199] W. K. Kwok, U. Welp, K. D. Carlson, G. W. Crabtree, K. G. Vandervoort, H. H. Wang, A. M. Kini, J. M. Williams, D. L. Stupka, L. K. Montgomery, and J. E. Thompson, Phys. Rev. B **42**, 8686 (1990)

[200] U. Welp, W. K. Kwok, G. W. Crabtree, K. G. Vandervoort, and J. Z. Liu, Phys. Rev. Lett. **62**, 1908 (1989); H. Claus, G. W. Crabtree, J. Z. Liu, W. K. Kwok, and A. Umezawa, J. Appl. Phys. **63**, 4170 (1988).

[201] H. Kobayashi, R. Kato, A. Kobayashi, S. Moriyama, Y. Nishio, K. Kajita, and W. Sasaki, Synth. Metals **27**, A283 (1988)

[202] T. P. Orlando, E. J. McNiff, Jr., S. Foner, and M. R. Beasley, Phys. Rev. B **19**, 4545 (1979)

[203] S. Wanka and J. Wosnitza, unpublished results

[204] M. V. Kartsovnik, V. M. Krasnov, and N. D. Kushch, Zh. Eksp. Teor. Fiz. **97**, 367 (1990) [Sov. Phys. JETP **70**, 208 (1990)]

[205] V. M. Krasnov, V. A. Larkin, and V. V. Ryazanov, Physica C **174**, 440 (1991); V. M. Krasnov, Physica C **190**, 357 (1992)

[206] H. Ito, M. Watanabe, Y. Nogami, T. Ishiguro, T. Komatsu, G. Saito, and N. Hosoito, J. Phys. Soc. Jpn. **60**, 3230 (1991); H. Ito, Y. Nogami, T. Ishiguro, T. Komatsu, G. Saito, and N. Hosoito, in *Mechanism of Superconductivity*, ed. by Y. Muto (Jpn. J. Appl. Phys. Publication Office, Tokyo 1992) Ser. 7, p. 419

[207] K. Oshima, H. Urayama, H. Yamochi, and G. Saito, J. Phys. Soc. Jpn, **57**, 730 (1988); K. Murata, Y. Honda, H. Anzai, M. Tokumoto, K. Takahashi, N. Kinoshita, T. Ishiguro, N. Toyota, T. Sasaki, and Y. Muto, Synth. Metals **27**, A341 (1988); H. Mori, K. Nakao, S. Nagaya, I. Hirabayashi, S. Tanaka, K. Oshima, and G. Saito, Synth. Metals **41–43**, 2159 (1991)

[208] G. R. Stewart, J. O'Rourke, G. W. Crabtree, K. D. Carlson, H. H. Wang, J. M. Williams, F. Gross, and K. Andres, Phys. Rev. B **33**, 2046 (1986)

[209] B. Andraka, C. S. Jee, J. S. Kim, G. R. Stewart, K. D. Carlson, H. H. Wang, A. V. S. Crouch, A. M. Kini, and J. M. Williams, Solid State Commun. **79**, 57 (1991)

[210] K. Andres, H. Schwenk, and H. Veith, Physica B **143**, 334 (1986)

[211] G. R. Stewart, J. M. Williams, H. H. Wang, L. N. Hall, M. T. Perozzo, and K. D. Carlson, Phys. Rev. B **34**, 6509 (1986); B. Andraka, J. S. Kim, G. R. Stewart, K. D. Carlson, H. H. Wang, and J. M. Williams, Phys. Rev. B **40**, 11345 (1989); B. Andraka, G. R. Stewart, K. D. Carlson, H. H. Wang, M. D. Vashon, and J. M. Williams, Phys. Rev. B **42**, 9963 (1990)

[212] J. E. Graebner, R. C. Haddon, S. V. Chichester, and S. H. Glarum, Phys. Rev. B **41**, 4808 (1990);

[213] K. I. Pohkhodnia, A. Graja, M. Weger, and D. Schweitzer, Z. Phys. B **90**, 127 (1993)

[214] B. Mühlschlegel, Z. Phys. **155**, 313 (1959)

[215] V. N. Kopylov and A. V. Palnichenko, J. Phys. I France **3**, 693 (1993)

[216] F. Creuzet, C. Bourbonnais, D. Jérome, D. Schweitzer, and H. J. Keller, Europhys. Lett. **1**, 467 (1986); F. Creuzet, C. Bourbonnais, G. Creuzet, D. Jérome, D. Schweitzer, and H. J. Keller, Physica B **143**, 363 (1986)

[217] Y. Maniwa, T. Takahashi, and G. Saito, J. Phys. Soc. Jpn. **55**, 47 (1986); Y. Maniwa, T. Takahashi, M. Takigawa, H. Yasuoka, G. Saito, K. Murata, M. Tokumoto, and H. Anzai, J. Phys. Soc. Jpn. **58**, 1048 (1989); T. Takahashi, T. Tokiwa, K. Kanoda, H. Urayama, H. Yamochi, and G. Saito, Physica C **153–155**, 487 (1988); Synth. Metals **27**, A319 (1988)

[218] S. M. DeSoto, C. P. Slichter, H. H. Wang, U. Geiser, and J. M. Williams, Phys. Rev. Lett. **70**, 2956 (1993)

[219] W. K. Kwok, U. Welp, V. M. Vinokur, S. Fleshler, J. Downey, and G. W. Crabtree, Phys. Rev. Lett. **67**, 390 (1991)

[220] K. Kanoda, K. Akiba, K. Suzuki, and T. Takahashi, Phys. Rev. Lett. **65**, 1271 (1990); T. Takahashi, K. Kanoda, K. Akiba, K. Sakao, M. Watabe, K. Suzuki, and G. Saito, Synth. Metals **41–43**, 2005 (1991)

[221] S. Sridhar, B. Maheswaran, B. A. Willemsen, D. H. Wu, and R. C. Haddon, Phys. Rev. Lett. **68**, 2220 (1992)

[222] L. P. Le, G. M. Luke, B. J. Sternlieb, W. D. Wu, Y. J. Uemura, J. H. Brewer, T. M. Riseman, C. E. Stronach, G. Saito, H. Yamochi, H. H. Wang, A. M. Kini, K. D. Carlson, and J. M. Williams, Phys. Rev. Lett. **68**, 1923 (1992)

[223] D. Achkir, M. Poirier, C. Bourbonnais, G. Quirion, C. Lenoir, P. Batail, and D. Jérome, Phys. Rev. B **47**, 11595 (1993)

[224] D. R. Harshman, R. N. Kleiman, R. C. Haddon, S. V. Chichester-Hicks, M. L. Kaplan, L. W. Rupp, Jr., T. Pfiz, D. L. Williams, and D. B. Mitzi, Phys. Rev. Lett. **64**, 1293 (1990); D. R. Harshman, A. T. Fiory, R. C. Haddon, M. L. Kaplan, T. Pfiz, E. Koster, I. Shinkoda, and D. L. Williams, Phys. Rev. B **49**, 12990 (1994)

[225] K. Holczer, D. Quinlivan, O. Klein, G. Grüner, and F. Wudl, Solid State Commun. **76**, 499 (1990); K. Holczer, O. Klein, G. Grüner, H. Yamochi, and F. Wudl, in *Organic Superconductivity*, ed. V. Z. Kresin and W. A. Little (Plenum, New York, 1990), p. 81; O. Klein, K. Holczer, G. Grüner, J. J. Chang, and F. Wudl, Phys. Rev. Lett. **66**, 655 (1991)

[226] M. Dressel, S. Bruder, G. Grüner, K. D. Carlson, H. H. Wang, and J. M. Williams, Phys. Rev. B **48**, 9906 (1993); M. Dressel, O. Klein, G. Grüner, K. D. Carlson, H. H. Wang, and J. M. Williams, Phys. Rev. B **50**, 13603 (1994)

[227] M. Lang, N. Toyota, T. Sasaki, and H. Sato, Phys. Rev. Lett. **69**, 1443 (1992); M. Lang, N. Toyota, and T. Sasaki, Physica B **186–188**, 1046 (1993)

[228] M. Lang, N. Toyota, T. Sasaki, and H. Sato, Phys. Rev. B **46**, 5822 (1992)

[229] G. Saito, H. Urayama, H. Yamochi, and K. Oshima, Synth. Metals **27**, A331 (1988)

[230] U. Geiser, A. M. Kini, H. H. Wang, J. A. Schlueter, K. D. Carlson, and J. M. Williams, unpublished results

[231] W. L. McMillan, Phys. Rev. **167**, 331 (1968)

[232] M. Tokumoto, H. Anzai, K. Murata, K. Kajimura, and T. Ishiguro, Jpn. J. Appl. Phys. **26**, Suppl. 26–3, 1977 (1987)

[233] Y. Hasegawa and H. Fukuyama, J. Phys. Soc. Jpn. **55**, 3717 (1986)

[234] A. Nowack, M. Weger, D. Schweitzer, and H. J. Keller, Solid State Commun. **60**, 199 (1986); A. Nowack, U. Poppe, M. Weger, D. Schweitzer, and H. Schwenk, Z. Phys. B **68**, 41 (1987); M. Weger, A. Nowack, and D. Schweitzer, Synth. Metals **41–43**, 1885 (1991); G. Ernst, A. Nowack, M. Weger, and D. Schweitzer, Europhys. Lett. **25**, 303 (1994)

[235] M. E. Hawley, K. E. Gray, B. D. Terris, H. H. Wang, K. D. Carlson, and J. M. Williams, Phys. Rev. Lett. **57**, 629 (1986)

[236] H. Bando, S. Kashiwaya, H. Tokumoto, H. Anzai, N. Kinoshita, and K. Kajimura, J. Vac. Sci. Technol. A **8**, 479 (1990)

[237] J. E. Schirber, D. L. Overmyer, K. D. Carlson, J. M. Williams, A. M. Kini, H. H. Wang, H. A. Charlier, B. J. Love, D. M. Watkins, and G. A. Yaconi, Phys. Rev. B **44**, 4666 (1991)

[238] Y. V. Sushko, H. Ito, T. Ishiguro, S. Horiuchi, and G. Saito, Solid State Commun. **87**, 997 (1993); Y. V. Sushko and K. Andres, Phys. Rev. B **47**, 330 (1993); Y. V. Sushko, H. Ito, T. Ishiguro, and G. Saito, Physica B **194–196**, 1999 & 2001 (1994)

[239] U. Welp, S. Fleshler, W. K. Kwok, G. W. Crabtree, K. D. Carlson, H. H. Wang, U. Geiser, J. M. Williams, and V. M. Hitsman, Phys. Rev. Lett. **69**, 840 (1992); Physica B **186–188**, 1065 (1993)

[240] K. Miyagawa, A. Kawamoto, Y. Nakazawa, and K. Kanoda, Phys. Rev. Lett. **75**, 1174 (1995)

[241] Y. V. Sushko, H. Ito, T. Ishiguro, S. Horuchi, and G. Saito, J. Phys. Soc. Jpn. **62**, 3372 (1993)

[242] H. Posselt, H. Müller, K. Andres, and G. Saito, Phys. Rev. B **49**, 15849 (1994); H. Posselt, K. Andres, G. Saito, and Y. V. Sushko, Solid State Commun. **92**, 613 (1994)

[243] Y. V. Sushko, K. Murata, H. Ito, T. Ishiguro, and G. Saito Synth. Metals **70**, 907 (1995)

[244] S. K. Sinha, G. W. Crabtree, D. G. Hinks, and H. Mook, Phys. Rev. Lett. **48**, 950 (1982)

[245] W. J. de Haas and P. M. van Alphen, Proc. Netherlands Roy. Acad. Sci. **33**, 680 & 1106 (1930)

[246] L. W. Schubnikow and W. J. de Haas, Proc. Netherlands Roy. Acad. Sci. **33**, 130 & 163 (1930).

[247] A. V. Gold in *Solid State Physics* Vol. 1, *Electrons in metals*, ed. J. F. Cochran and R. R. Haering (Simon Fraser University Lectures 1986) p. 39; J. M. Ziman (ed.) *The Physics of Metals 1. Electrons* (Cambridge University Press 1969); M. Springford (ed.) *Electrons at the Fermi Surface* (Cambridge University Press 1980)

[248] J. R. Anderson and D. R. Stone in *Methods of Experimental Physics* Vol. 11, ed. R. V. Coleman (Academic Press, New York 1974), p. 33

[249] D. Shoenberg, *Magnetic Oscillations in Metals*, (Cambridge University Press, Cambridge 1984)

[250] L. D. Landau in Appendix of D. Shoenberg, Proc. Roy. Soc. London Ser. A **170**, 341 (1939)

[251] I. M. Lifshitz and A. M. Kosevich, Zh. eksp. teor. fiz. **29**, 730 (1955) [Sov. Phys. JETP **2**, 636 (1956)]

[252] R. B. Dingle, Proc. Roy. Soc. London Ser. A **211**, 517 (1952)

[253] D. Pines and P. Nozières, *The Theory of Quantum Liquids* (Benjamin, New York 1966)

[254] S. Engelsberg and G. Simpson, Phys. Rev. B **2**, 1657 (1970); S. Engelsberg, Phys. Rev. B **18**, 966 (1987); P. S. Riseborough, Phys. Status Solidi B **122**, 161 (1984)

[255] J. S. Brooks, M. J. Naughton, Y. P. Ma, P. M. Chaikin, and R. V. Chamberlin, Rev. Sci. Instrum. **58**, 117 (1987)

[256] E. N. Adams and T. D. Holstein, J. Phys. Chem. Solids **10**, 254 (1959)

[257] A. B. Pippard, *The Dynamics of Conduction Electrons* (Gordon and Breach, New York 1965)

[258] K. Kajita, Y. Nishio, T. Takahashi, W. Sasaki, R. Kato, H. Kobayashi, A. Kobayashi, and Y. Iye, Solid State Commun. **70**, 1189 (1989)

[259] V. G. Paschansky, J. A. Roldan Lopez, and T. G. Yao, J. Phys. I France **1**, 1469 (1991)

[260] T. Osada, A. Kawasumi, S. Kagoshima, N. Miura, and G. Saito, Phys. Rev. Lett. **66**, 1525 (1991); Physica C **185–189**, 2697 (1991); M. J. Naughton, O. H. Chung, M. Chaparta, X. Bu, and P. Coppens, Phys. Rev. Lett. **67**, 3712 (1991)

[261] T. Osada, S. Kagoshima, and N. Miura, Phys. Rev. B **46**, 1812 (1992)

[262] A. G. Lebed', Pis'ma Zh. Eksp. Teor. Fiz. **43**, 137 (1986) [JETP Lett. **43**, 174 (1986)]

[263] W. Kang, S. T. Hannahs, and P. M. Chaikin, Phys. Rev. Lett. **69**, 2827 (1992); K. Behnia, M. Ribault, and C. Lenoir, Europhys. Lett. **25**, 285 (1994)

[264] A. G. Lebed' and P. Bak, Phys. Rev. Lett. **63**, 1315 (1989)

[265] K. Maki, Phys. Rev. B **45**, 5111 (1992)

[266] P. M. Chaikin, Phys. Rev. Lett. **69**, 2831 (1992)

[267] A. T. Zheleznyak and V. M. Yakovenko, Synth. Metals **70**, 1005 (1995)

[268] S. P. Strong, D. G. Clarke, and P. W. Anderson, Phys. Rev. Lett. **73**, 1007 (1994)

[269] D. G. Clarke, S. P. Strong, and P. W. Anderson, Phys. Rev. Lett. **72**, 3218 (1994)

[270] M. V. Kartsovnik, A. E. Kovalev, V. N. Laukhin, and S. I. Pesotskii, J. Phys. I France **2**, 223 (1992)

[271] G. M. Danner, W. Kang, and P. M. Chaikin, Phys. Rev. Lett. **72**, 3714 (1994); Physica B **201**, 442 (1994)

[272] W. Kang, J. R. Cooper, and D. Jérome, Phys. Rev. B **43**, 11467 (1991)

[273] R. W. Stark and C. B. Friedberg, J. Low. Temp. Phys. **14**, 111 (1974)

[274] E. I. Blount, Phys. Rev. **126**, 1636 (1962)

[275] Y. Aharonov and D. Bohm, Phys. Rev. **115**, 485 (1959)

[276] G. O. Baram, L. I. Buravov, L. C. Degtyarev, M. É. Kozlov, V. N. Laukhin, E. É. Laukhina, V. G. Onishchenko, K. I. Pokhodnya, M. K. Scheinkman, R. P. Shibaeva, and É. B. Yagubskii, Pis'ma Zh. Eksp. Teor. Fiz. **44**, 293 (1986) [JETP Lett. **44**, 376 (1986)]; D. Schweitzer, P. Bele, H. Brunner, E. Gogu, U. Haeberlen, I. Hennig, I. Klutz, R. Świetlik, and H. J. Keller, Z. Phys. B **67**, 489 (1987)

[277] H. Ito, H. Kaneko, T. Ishiguro, H. Ishimoto, K. Kono, S. Horiuchi, T. Komatsu, and G. Saito, Solid State Commun. **85**, 1005 (1993); H. Ito, T. Ishiguro, H. Ishimoto, K. Kono, M. V. Kartsovnik, H. Mori, S. Tanaka, G. Saito, and N. D. Kushch, to be published

[278] J. Wosnitza, G. W. Crabtree, H. H. Wang, K. D. Carlson, M. D. Vashon, and J. M. Williams, Phys. Rev. Lett. **67**, 263 (1991)

[279] J. Wosnitza, G. W. Crabtree, H. H. Wang, U. Geiser, J. M. Williams, and K. D. Carlson, Phys. Rev. B **45**, 3018 (1992)

[280] T. Osada, A. Kawasumi, R. Yagi, S. Kagoshima, N. Miura, M. Oshima, H. Mori, T. Nakamura, and G. Saito, Solid State Commun. **75**, 901 (1990); J. Singleton, F. L. Pratt, M. Doporto, J. Caulfield, W. Hayes, I. Deckers, G. Pitsi, F. Herlach, T. J. B. M. Janssen, J. A. A. J. Perenboom, M. Kurmoo, and P. Day, Synth. Metals **55–57**, 2198 (1993)

[281] D. L. Randles, Proc. Roy. Soc. London Ser. A **331**, 85 (1972); G. W. Crabtree, L. R. Windmiller, and J. B. Ketterson, J. Low Temp. Phys. **20**, 655 (1975)

[282] S. Hill, J. Singleton, F. L. Pratt, M. Doporto, W. Hayes, T. J. B. M. Janssen, J. A. A. J. Perenboom, M. Kurmoo, and P. Day, Synth. Metals **55–57**, 2566 (1993); J. Singleton, F. L. Pratt, M. Doporto, J. M. Caulfield, S. O. Hill, T. J. B. M. Janssen, I. Dekkers, G. Pitsi, F. Herlach, W. Hayes, J. A. A. J. Perenboom, M.Kurmoo, and P. Day, Physica B **184**, 470 (1993); S. Hill, A. Wittlin, J. van Bentum, J. Singleton, W. Hayes, J. A. A. J. Perenboom, M. Kurmoo, and P. Day, Synth. Metals **70**, 821 (1995)

[283] W. Kohn, Phys. Rev. **123**, 1242 (1961)

[284] N. Toyota, E. W. Fenton, T. Sasaki, and M. Tachiki, Solid State Commun. **72**, 859 (1989); J. Singleton, J. Caulfield, S. Hill, S. Blundell, W. Lubcynski, A. House, W. Hayes, J. Perenboom, M. Kurmoo, and P. Day, Physica B **211**, 275 (1995)

[285] A. E. Kovalev, M. V. Kartsovnik, and N. D. Kushch, Solid State Commun. **87**, 705 (1993); M. V. Kartsovnik, D. V. Mashovets, D. V. Smirnov, V. N. Laukhin, A. Gilewski, and N. D. Kushsh, J. Phys. I France **4**, 159 (1994); M. V. Kartsovnik, A. E. Kovalev, D. V. Mashovets, D. V. Smirnov, V. N. Laukhin, A. Gilewskii, and N. D. Kushch, Physica B **201**, 463 (1994)

[286] N. Herrmann and J. Wosnitza, unpublished results

[287] G. Goll and J. Wosnitza, to be published

[288] V. N. Laukhin, A. Audouard, H. Rakoto, J. M. Broto, F. Goze, G. Coffe, L. Brossard, J. P. Redoules, M. V. Kartsovnik, N. D. Kushch, L. I. Buravov, A. G. Khomenko, É. B. Yagubskii, S. Askenazy, and P.Pari, Physica B **211**, 282 (1995)

[289] T. Osada, R. Yagi, A. Kawasumi, S. Kagoshima, N. Miura, M. Oshima, and G. Saito, Phys. Rev. B **41**, 5428 (1990); T. Osada, R. Yagi, A. Kawasumi, S. Kagoshima, N. Miura, M. Oshima, H. Mori, T. Nakamura, and G. Saito, Synth. Metals **41–43**, 2171 (1991)

[290] J. Caulfield, S. J. Blundell, M. S. L. du Croo de Jongh, P. T. J. Hendriks, J. Singleton, M. Doporto, F. L. Pratt, A. House, J. A. A. J. Perenboom, W. Hayes, M. Kurmoo, and P. Day, Phys. Rev. B **51**, 8325 (1995)

[291] N. Harrison, A. House, I. Deckers, J. Caulfield, J. Singleton, F. Herlach, W. Hayes, M. Kurmoo, and P. Day, Phys. Rev. B **52**, 5584 (1995)

[292] P. Christ, W. Biberacher, H. Müller, and K. Andres, Solid State Commun. **91**, 451 (1994)

[293] J. Caulfield and J. Singleton, private communication

[294] T. Sasaki and N. Toyota, Solid State Commun. **82**, 447 (1992); Synth. Metals **55–57**, 2296 (1993)

[295] J. S. Brooks, C. C. Agosta, S. J. Klepper, M. Tokumoto, N. Kinoshita, H. Anzai, S. Uji, H. Aoki, A. S. Perel, G. J. Athas, and D. A. Howe, Phys. Rev. Lett. **69**, 156 (1992)

[296] T. Sasaki, H. Sato, and N. Toyota, Synth. Metals **41–43**, 2211 (1991)

[297] N. Kinoshita, M. Tokumoto, and H. Anzai, J. Phys. Soc. **59**, 3410 (1990)

[298] M. Tokumoto, A. G. Swanson, J. S. Brooks, C. C. Agosta, S. T. Hannahs, N. Kinoshita, H. Anzai, M. Tamura, H. Tajima, H. Kuroda, and J. R. Anderson, in *Organic Superconductivity*, ed. V. Z. Kresin and W. A. Little (Plenum, New York, 1990), p. 167; M. Tokumoto, N. Kinoshita, H. Anzai, A. G. Swanson, J. S. Brooks, S. T. Hannahs, C. C. Agosta, M. Tamura, H. Tajima, H. Kuroda, A. Ugawa, and K. Yakushi, Synth. Metals **41–43**, 2459 (1991)

[299] M. Tokumoto, A. G. Swanson, J. S. Brooks, C. C. Agosta, S. T. Hannahs, N. Kinoshita, H. Anzai, and J. R. Anderson, J. Phys. Soc. Jpn. **59**, 2324 (1990); T. Sasaki, N. Toyota, M. Tokumoto, N. Kinoshita, and H. Anzai, Solid State Commun. **75**, 97 (1990)

[300] F. L. Pratt, M. Doporto, J. Singleton, T. J. B. M. Janssen, J. A. A. J. Perenboom, M. Kurmoo, W. Hayes, and P. Day, Physica B **177**, 333 (1992)

[301] F. L. Pratt, J. Singleton, M. Doporto, A. J. Fisher, T. J. B. M. Janssen, J. A. A. J. Perenboom, M. Kurmoo, W. Hayes, and P. Day, Phys. Rev. B **45**, 13904 (1992); W. Biberacher, C.-P. Heidmann, H. Müller, W. Joss, Ch. Probst, and K. Andres, Synth. Metals **55–57**, 2241 (1993)

[302] T. Sasaki and N. Toyota, Phys. Rev. B **48**, 11457 (1993)

[303] G. J. Athas, J. S. Brooks, S. Valfells, S. J. Klepper, M. Tokumoto, N. Kinoshita, T. Kinoshita, and Y. Tanaka, Phys. Rev. B **50**, 17713 (1994)

[304] S. Uji, H. Aoki, J. S. Brooks, A. S. Perel, G. J. Athas, S. J. Klepper, C. C. Agosta, D. A. Howe, M. Tokumoto, N. Kinoshita, Y. Tanaka, and H. Anzai, Solid State Commun. **88**, 683 (1993)

[305] M. V. Kartsovnik, A. E. Kovalev, V. N. Laukhin, S. I. Pesotskii, and N. D. Kushch, Pis'ma Zh. Eksp. Teor. Fiz. **55**, 337 (1992) [JETP Lett. **55**, 339 (1992)]

[306] S. Uji, H. Aoki, M. Tokumoto, T. Kinoshita, N. Kinoshita, Y. Tanaka, and H. Anzai, Phys. Rev. B **49**, 732 (1994); S. Uji, T. Terashima, H. Aoki, M. Tokumoto, T. Kinoshita, N. Kinoshita, Y. Tanaka, and H. Anzai, Physica B **210**, 479 (1994); S. Uji, T. Terashima, H. Aoki, J. S. Brooks, M. Tokumoto, N. Kinoshita, T. Kinoshita, Y. Tanaka, and H. Anzai, Synth. Metals **70**, 807 (1995); S. Uji, T. Terashima, H. Aoki, J. S. Brooks, M. Tokumoto, N. Kinoshita, T. Kinoshita, Y. Tanaka, and H. Anzai, J. Phys.: Condens. Matter **6**, L539 (1994)

[307] J. S. Brooks, S. J. Klepper, C. C. Agosta, M. Tokumoto, N. Kinoshita, Y. Tanaka, S. Uji, H. Aoki, G. J. Athas, X. Chen, and H. Anzai, Synth. Metals **55–57**, 1791 (1993)

[308] S. J. Klepper, J. S. Brooks, G. J. Athas, X. Chen, M. Tokumoto, N. Kinoshita, and Y. Tanaka, Physica B **194–196**, 2003 (1994)

[309] Y. Iye, R. Yagi, N. Hanasaki, S. Kagoshima, H. Mori, H. Fujimoto, and G. Saito, J. Phys. Soc. Jpn. **63**, 674 (1994); Physica B **201**, 474 (1994); T. Sasaki and N. Toyota, Phys. Rev. B **49**, 10120 (1994); Physica B **201**, 466 (1994); Synth. Metals **70**, 849 (1995)

[310] J. Caulfield, J. Singleton, P. T. J. Hendriks, J. A. A. J. Perenboom, F. L. Pratt, M. Doporto, W. Hayes, M. Kurmoo, and P. Day, J. Phys.: Condens. Matter **6**, L155 (1994)

[311] A. E. Kovalev, M. V. Kartsovnik, R. P. Shibaeva, L. P. Rozenberg, I. F. Schegolev, and N. D. Kushch, Solid State Commun. **89**, 575 (1994); M.

V. Kartsovnik, A. E. Kovalev, R. P. Shibaeva, L. P. Rozenberg, and N. D. Kushch, Physica B **201**, 459 (1994)

[312] M. V. Kartsovnik, H. Ito, T. Ishiguro, H. Mori, T. Mori, G. Saito, and S. Tanaka, J. Phys.: Condens. Matter **6**, L479 (1994)

[313] M. V. Kartsovnik, A. E. Kovalev, and N. D. Kushch, J. Phys. I France **3**, 1187 (1993)

[314] S. J. Klepper, J. S. Brooks, X. Chen, I. Bradaric, M. Tokumoto, N. Kinoshita, Y. Tanaka, and C. C. Agosta, Phys. Rev. B **48**, 9913 (1993)

[315] C. E. Campos, J. S. Brooks, P. J. M. van Bentum, J. A. A. J. Perenboom, S. J. Klepper, P. S. Sandhu, M. Tokumoto, T. Kinoshita, N. Kinoshita, Y. Tanaka, and H. Anzai, Synth. Metals **70**, 803 (1995)

[316] C. E. Campos, J. S. Brooks, P. J. M. van Bentum, J. A. A. J. Perenboom, S. J. Klepper, P. S. Sandhu, S. Valfells, Y. Tanaka, T. Kinoshita, N. Kinoshita, M. Tokumoto, and H. Anzai, Physica B **211**, 293 (1995)

[317] M. Doporto, F. L. Pratt, J. Singleton, M. Kurmoo, and W. Hayes, Phys. Rev. Lett. **69**, 991 (1992)

[318] J. Singleton, F. L. Pratt, M. Doporto, T. J. B. M. Janssen, M. Kurmoo, J. A. A. J. Perenboom, W. Hayes, and P. Day, Phys. Rev. Lett. **68**, 2500 (1992)

[319] K. Murata, N. Toyota, Y. Honda, T. Sasaki, M. Tokumoto, H. Bando, H. Anzai, Y. Muto, and T. Ishiguro, J. Phys. Soc. Jpn. **57**, 1540 (1988); N. Toyota, T. Sasaki, K. Murata, Y. Honda, M. Tokumoto, H. Bando, N. Kinoshita, H. Anzai, T. Ishiguro, and Y. Muto, J. Phys. Soc. Jpn. **57**, 2616 (1988)

[320] M. V. Kartsovnik, P. A. Kononovich, V. N. Laukhin, S. I. Pesotskii, and I. F. Schegolev, Zh. Eksp. Teor. Fiz. **49**, 453 (1989) [JETP Lett. **49**, 519 (1989)]

[321] V. N. Laukhin, M. V. Kartsovnik, S. I. Pesotskii, P. A. Kononovich, and I. F. Schegolev, Synth. Metals **41–43**, 2453 (1991)

[322] M. V. Kartsovnik, P. A. Kononovich, V. N. Laukhin, S. I. Pesotskii, and I. F. Schegolev, Zh. Eksp. Teor. Fiz. **97**, 1305 (1990) [Sov. Phys. JETP **70**, 735 (1990)]; M. V. Kartsovnik, V. N. Laukhin, and S. I. Pesotskii, Fiz. Nizk. Temp. **18**, 22 (1992) [Sov. J. Low Temp. Phys. **18**, 13 (1992)]

[323] H. Endres, H. J. Keller, R. Swietlik, D. Schweitzer, K. Angermund, and C. Krüger, Z. Naturforsch. **41** a 1319 (1986)

[324] W. Kang, G. Montambaux, J. R. Cooper, D. Jérome, P. Batail, and C. Lenoir, Phys. Rev. Lett. **62**, 2559 (1989)

[325] J. Wosnitza, D. Beckmann, S. Wanka, and D. Schweitzer, unpublished results

[326] I. D. Vagner, T. Maniv, and E. Ehrenfreund, Phys. Rev. Lett. **51**, 1700 (1983); K. Jauregi, V. I. Marchenko, and I. D. Vagner, Phys. Rev. B **41**, 12922 (1990)

[327] J. H. Condon, Phys. Rev. **145**, 526 (1966)

[328] J. Wosnitza, G. W. Crabtree, H. H. Wang, K. D. Carlson, and J. M. Williams, in *Physical Phenomena at High Magnetic Fields*, ed. E. Manousakis, P. Schlottmann, P. Kumar, K. Bedell, and F. M. Mueller (Addison–Wesley, Redwood City, CA, 1992), p. 411; J. Wosnitza, G. W. Crabtree, J. M. Williams, H. H. Wang, K. D. Carlson, and U. Geiser, Synth. Metals **55–57**, 2891 (1993); J.Wosnitza, G. W. Crabtree, K. D. Carlson, H. H. Wang, and J. M. Williams, Physica B **194–196**, 2007 (1994)

[329] J. Wosnitza, N. Herrmann, unpublished results

[330] T. Sugano, G. Saito, and M. Kinoshita, Phys. Rev. B **35**, 6554 (1987)

[331] I. D. Parker, D. D. Pigram, R. H. Friend, M. Kurmoo, and P. Day, Synth. Metals **27**, A387 (1988)

[332] K. Kajita, Y. Nishio, S. Moriyama, W. Sasaki, R. Koto, H. Kobayashi, and A. Kobayashi, Solid State Commun. **60**, 811 (1986)

[333] M. Kurmoo, D. R. Talham, P. Day, I. D. Parker, R. H. Friend, A. M. Stringer, and J. A. K. Howard, Solid State Commun. **61**, 459 (1987)

[334] S. Uji, H. Aoki, M. Tokumoto, A. Ugawa, and K. Yakushi, Physica B **194–196**, 1307 (1994); Mat. Res. Soc. Symp. Proc. Vol. 328, 337 (1994)

[335] T. Mori, F. Sakai, G. Saito, and H. Inokuchi, Chem. Lett. **1986**, 1037 (1986)

[336] M. Doporto, J. Singleton, F. L. Pratt, J. Caulfield, W. Hayes, J. A. A. J. Perenboom, I. Deckers, G. Pitsi, M. Kurmoo, and P. Day, Phys. Rev. B **49**, 3934 (1994)

[337] M. Doporto, J. Caulfield, F. L. Pratt, J. Singleton, S. Hill, W. Hayes, J. A. A. J. Perenboom, M. Kurmoo, and P. Day, Synth. Metals **55–57**, 2572 (1993); F. L. Pratt, M. Doporto, J. Singleton, T. J. B. M. Janssen, J. A. A. J. Perenboom, M. Kurmoo, W. Hayes, and P. Day, Physica B **177**, 333 (1992)

[338] F. L. Pratt, A. J. Fisher, W. Hayes, J. Singleton, S. J. R. M. Spermon, M. Kurmoo, and P. Day, Phys Rev. Lett. **61**, 2721 (1988)

[339] M. Doporto, F. L. Pratt, W. Hayes, J. Singleton, T. Janssen, M. Kurmoo, and P. Day, Synth. Metals **41–43**, 1903 (1991)

[340] A. G. Swanson, J. S. Brooks, M. Tokumoto, A. Ugawa, and K. Yakushi, in *Organic Superconductivity*, ed. V. Z. Kresin and W. A. Little (Plenum, New York, 1990), p. 191

[341] M. V. Kartsovnik, G. Yu. Logvenov, H. Ito, T. Ishiguro, and G. Saito, Phys. Rev. B **52**, 15715 (1995); M. V. Kartsovnik, private communication

[342] Y. C. Jean, Y. Lou, H. L. Yen, K. M. O'Brien, R. N. West, H. H. Wang, K. D. Carlson, and J. M. Williams, Physica C **221**, 399 (1994)

[343] K. Oshima, T. Mori, H. Inokuchi, H. Urayama, H. Yamochi, and G. Saito, Synth. Metals **27**, A165 (1988)

[344] M. V. Kartsovnik, P. A. Kononovich, V. N. Laukhin, R. B. Lyubovskii, and S. I. Pesotskii, Zh. Eksp. Teor. Fiz. **98**, 708 (1990) [Sov. Phys. JETP **71**, 396 (1990)]

[345] N. Toyota, T. Sasaki, K. Murata, Y. Honda, M. Tokumoto, H. Bando, N. Kinoshita, H. Anzai, T. Ishiguro, and Y. Muto, J. Phys. Soc. Jpn. **57**, 2616 (1988); T. Sasaki, H. Sato, and N. Toyota, Solid State Commun. **76**, 507 (1990)

[346] A. G. Swanson, J. S. Brooks, H. Anzai, N. Kinoshita, M. Tokumoto, and K. Murata, Solid State Commun. **73**, 353 (1990)

[347] C.-P. Heidmann, W. Biberacher, H. Müller, W. Joss, and K. Andres, in: *Lower-Dimensional Systems and Molecular Electronics*, ed. R. M. Metzger, P. Day, and G. C. Papavassiliou, NATO ASI Ser. B, Vol. 248 (Plenum, New York 1991) p. 233; K. Andres, C.-P. Heidmann, H. Müller, S. Himmelsbacher, C. Probst, and W. Joss, Synth. Metals **41–43**, 1893 (1991); C.-P. Heidmann, H. Müller, W. Biberacher, K. Neumaier, C. Probst, K. Andres, A. G. M. Jansen, and W. Joss, Synth. Metals **41–43**, 2029 (1991)

[348] F. L. Pratt, J. Singleton, M. Kurmoo, S. J. R. M. Spermon, W. Hayes, and P. Day, in: *The Physics and Chemistry of Organic Superconductors*, ed. G. Saito and S. Kagoshima (Springer, Berlin, Heidelberg, 1990), p. 200; F. L. Pratt, M. Doporto, W. Hayes, J. Singleton, T. Janssen, M. Kurmoo, and P. Day, Synth. Metals **41–43**, 2195 (1991)

[349] J. Caulfield, W. Lubczynski, F. L. Pratt, J. Singleton, D. Y. K. Ko, W. Hayes, M. Kurmoo, and P. Day, J. Phys.: Condens. Matter **6**, 2911 (1994); W. Lubczynski, J. Caulfield, J. Singleton, F. L. Pratt, A. House, W. Hayes, M. Kurmoo, and P. Day, Physica C **235–240**, 2457 (1994); W. Lubczynski, J. Caulfield, F. L. Pratt, J. Singleton, W. Hayes, M. Kurmoo, and P. Day, Physica B **201**, 483 (1994)

[350] N. Toyota and W. Sasaki, Physica B **186–188**, 1056 (1993); T. Sasaki and N. Toyota, Synth. Metals **55–57**, 2303 (1993)

[351] F. A. Meyer, E. Steep, W. Biberacher, P. Christ, A. Lerf, A. G. M. Jansen, W. Joss, P. Wyder, and K. Andres, Europhys. Lett. **32**, 681 (1995)

[352] M. V. Kartsovnik, A. E. Kovalev, V. N. Laukhin, H. Ito, T. Ishiguro, N. D. Kushch, H. Anzai, and G. Saito, Synth. Metals **70**, 819 (1995)

[353] J. E. Graebner and M. Robbins, Phys. Rev. Lett. **36**, 433 (1976); R. Corcoran, P. Meeson, Y. Onuki, P.-A. Probst, M. Springford, and K. Takita, Physica B **194–196**, 1573 (1994)

[354] R. Corcoran, N. Harrison, S. M. Hayden, P. Messon, M. Springford, and P. J. van der Wel, Phys. Rev. Lett. **72**, 701 (1994)

[355] N. Harrison, S. M. Hayden, P. Meeson, M. Springford, P. J. van der Wel, and A. A. Menovsky, Phys. Rev. B, to be published

[356] M. Heinecke and K. Winzer, Z. Phys. B **98**, 147 (1995)

[357] G. Goll, M. Heinecke, A. G. M. Jansen, W. Joss, L. Nguyen, E. Steep, K. Winzer, and P. Wyder, Phys. Rev. B, to be published

[358] K. Maki, Phys. Rev. B **44**, 2861 (1991); A. Wasserman and M. Springford, Physica B **194–196**, 1801 (1994)

[359] P. Miller and B. L. Györffy, J. Phys.: Condens. Matter **7**, 5579 (1995)

[360] L. P. Chan, K. G. Lynn, D. R. Harshman, and R. C. Haddon, Phys. Rev. B **50**, 10393 (1994)

[361] K. Oshima, H. Yamazaki, K. Kato, Y. Maruyama, R. Kato, A. Kobayashi, and H. Kobayashi, Synth. Metals **55–57**, 2334 (1993)

[362] M. Heinecke, K. Winzer, and D. Schweitzer, Z. Phys. B **93**, 45 (1993)

[363] E. Balthes, D. Schweitzer, I. Heinen, H. J. Keller, W. Biberacher, A. G. M. Jansen, and E. Steep, Synth. Metals **70**, 841 (1995); E. Balthes, D. Schweitzer, I. Heinen, H. J. Keller, W. Strunz, W. Biberacher, A. G. M. Jansen, and E. Steep, Z. Phys. B, to be published

[364] E. Balthes and D. Schweitzer, private communication; D. Beckmann and J. Wosnitza, unpublished results

[365] F. Wilczek, Phys. Rev. Lett. **49**, 957 (1982)

[366] U. G. L. Lahaise, Q. Chen, L. E. De Long, C. P. Brock, H. H. Wang, K. D. Carlson, J. A. Schlueter, and J. M. Williams, Phys. Rev. B **51**, 3301 (1995)

[367] H. Mori, I. Hirabayashi, S. Tanaka, T. Mori, and H. Inokuchi, Solid State Commun. **76**, 35 (1990)

[368] K. Oshima, K. Araki, H. Yamazaki, K. Kato, Y. Maruyama, K. Yakushi, T. Mori, H. Inokuchi, H. Mori, and S. Tanaka, Physica C **185–189**, 2689 (1991); K. Oshima, T. Doi, Y. Tokuoka, H. Yamazaki, K. Kato, Y. Maruyama, H. Mori, and S. Tanaka, Synth. Metals **55–57**, 2339 (1993)

[369] K. Kanoda, K. Kato, A. Kawamoto, K. Oshima, T. Takahashi, K. Kikuchi, K. Saito, and I. Ihemoto, Synth. Metals **55–57**, 2309 (1993)

[370] K. Kajita, Y. Nishio, T. Takahashi, W. Sasaki, R. Kato, H. Kobayashi, and A. Kobayashi, Solid State Commun. **70**, 1181 (1989)

[371] M. Tamura, H. Tajima, H. Kuroda, and M. Tokumoto, J. Phys. Soc. Jpn. **59**, 1753 (1990)

[372] M. Tokumoto, A. G. Swanson, J. S. Brooks, M. Tamura, H. Tajima, and H. Kuroda, Solid State Commun. **75**, 439 (1990); M. Tokumoto, A. G. Swanson, J. S. Brooks, C. C. Agosta, S. T. Hannahs, N. Kinoshita, H. Anzai, M. Tamura, H. Tajima, H. Kuroda, A. Ugawa, and K. Yakushi, Physica B **184**, 508 (1993)

[373] T. Terashima, S. Uji, H. Aoki, M. Tamura, M. Kinoshita, and M. Tokumoto, Solid State Commun. **91**, 595 (1994); Synth. Metals **70**, 847 (1995)

[374] M. Tamura, K. Yakushi, H. Kuroda, A. Kobayashi, R. Kato, and H. Kobayashi, J. Phys. Soc. Jpn. **57**, 3239 (1988)

[375] M. Tamura, H. Kuroda, S. Uji, H. Aoki, M. Tokumoto, A. G. Swanson, J. S. Brooks, C. C. Agosta, and S. T. Hannahs, J. Phys. Soc. Jpn. **63**, 615 (1994)

[376] T. Terashima, S. Uji, H. Aoki, M. Tamura, M. Kinoshita, and M. Tokumoto, Synth. Metals **70**, 845 (1995)

[377] S. J. Klepper, G. J. Athas, J. S. Brooks, M. Tokumoto, T. Kinoshita, N. Tamura, and M. Kinoshita, Synth. Metals **70**, 835 (1995)

[378] S. Kahlich, D. Schweitzer, P. Auban Senzier, D. Jérome, and H. J. Keller, Solid State Commun. **83**, 77 (1992); Synth. Metals **55–57**, 2483 (1993)

[379] S. Kahlich, D. Schweitzer, C. Rovira, J. A. Paradis, M.-H. Whangbo, I. Heinen, H. J. Keller, B. Nuber, P. Bele, H. Brunner, and R. P. Shibaeva, Z. Phys. B **94**, 39 (1994)

[380] H. Tajima, S. Ikeda, M. Inokuchi, A. Kobayashi, T. Ohta, T. Sasaki, N. Toyota, R. Kato, H. Kobayashi, and H. Kuroda, Solid State Commun. **88**, 605 (1993); H. Tajima, S. Ikeda, M. Inokuchi, T. Ohta, A. Kobayashi, T. Sasaki, N. Toyota, R. Kato, H. Kobayashi, and H. Kuroda, Synth. Metals **70**, 1051 (1995)

[381] A. Kobayashi, T. Naito, and H. Kobayashi, Synth. Metals **70**, 1047 (1995)

[382] H. Tajima, S. Ikeda, A. Kobayashi, H. Kuroda, R. Kato, and H. Kobayashi, Solid State Commun. **82**, 157 (1992); Synth. Metals **55–57**, 2530 (1993)

[383] R. Kato, H. Kobayashi, H. Kim, A. Kobayashi, Y. Sasaki, T. Mori, and H. Inokuchi, Chem. Lett. **1988**, 865 (1988)

[384] A. Aumüller, P. Erk, G. Klebe, S. Hünig, J. U. von Schütz, and H.-P. Werner, Angew. Chem. Int. Ed. Engl. **25**, 740 (1986); R. Kato, H. Kobayashi, A. Kobayashi, T. Mori, and H. Inokuchi, Chem. Lett. **1987**, 1579 (1987)

[385] S. Hünig, K. Sinzger, M. Jopp, D. Bauer, W. Bietsch, J. U. von Schütz, and H. C. Wolf, Angew. Chem. Int. Ed. Engl. **31**, 859 (1992); K. Sinzger, S. Hünig, M. Jopp, D. Bauer, W. Bietsch, J. U. von Schütz, H. C. Wolf, R. K. Kremer, T. Metzenthin, R. Bau, S. I. Khan, A. Lindbaum, C. L. Lengauer, and E. Tillmanns, J. Am. Chem. Soc. **115**, 7696 (1993)

[386] I. H. Inoue, A. Kakizaki, H. Namatame, A. Fujimori, A. Kobayashi, R. Kato, and H. Kobayashi, Phys. Rev. B **45**, 5828 (1992)

[387] K. Yakushi, A. Ugawa, G. Ojima, T. Ida, H. Tajima, H. Kuroda, R. Kato, and H. Kobayashi, Mol. Cryst. Liq. Cryst. **181**, 217 (1990)

[388] S. Uji, T. Terashima, H. Aoki, J. S. Brooks, R. Kato, H. Sawa, S. Aonuma, M. Tamura, and M. Kinoshita, Phys. Rev. B **50**, 15597 (1994); Synth. Metals **70**, 1075 (1995)

[389] T. Miyazaki, K. Terakura, Y. Morikawa, and T. Yamasaki, Phys. Rev. Lett. **74**, 5104 (1995)

[390] S. Uji, T. Terashima, H. Aoki, R. Kato, H. Sawa, S. Aonuma, M. Tamura, and M. Kinoshita, Solid State Commun. **93**, 203 (1995)

[391] Y. Nishio, K. Kajita, W. Sasaki, R. Kato, A. Kobayashi, and H. Kobayashi, Solid State Commun. **81**, 473 (1992)

[392] H. Fukuyama, J. Phys. Soc. Jpn. **61**, 3452 (1992); Y. Suzumura and Y. Ono, J. Phys. Soc. Jpn. **62**, 3244 (1993); M. Kato, M. Nakano, and K. Yamada, Physica B **194–196**, 1257 (1994)

Index

Springer Tracts in Modern Physics